Electrospun Polymer Nanofibers

Electrospun Polymer Nanofibers

Arkadii Arinstein

Pan Stanford Publishing

Published by

Pan Stanford Publishing Pte. Ltd.
Penthouse Level, Suntec Tower 3
8 Temasek Boulevard
Singapore 038988

Email: editorial@panstanford.com
Web: www.panstanford.com

British Library Cataloguing-in-Publication Data
A catalogue record for this book is available from the British Library.

Electrospun Polymer Nanofibers

ISBN 978-981-4745-27-7 (Hardcover)
ISBN 978-1-315-36462-9 (eBook)

Printed in the USA

Electrospun Polymer Nanofibers

Arkadii Arinstein

PAN STANFORD PUBLISHING

Published by

Pan Stanford Publishing Pte. Ltd.
Penthouse Level, Suntec Tower 3
8 Temasek Boulevard
Singapore 038988

Email: editorial@panstanford.com
Web: www.panstanford.com

British Library Cataloguing-in-Publication Data
A catalogue record for this book is available from the British Library.

Electrospun Polymer Nanofibers

ISBN 978-981-4745-27-7 (Hardcover)
ISBN 978-1-315-36462-9 (eBook)

Printed in the USA

Contents

Preface

"By making something absolutely clear, someone will become confused."

—Arthur Bloch

Polymer nanofibers fabricated through electrospinning have attracted much attention recently because of their unique mechanical and thermodynamic properties compared to the bulk.

Despite the prehistory of electrospinning spanning more than 100 years, the real examination of this process and its adaptation to the praxis started only 15–20 years ago. Nowadays, the hydrodynamics of the electrospinning process is well understood, which has enabled fabrication of polymer nanofibers in a regular and stable way, and electrospun polymer nanofibers are utilized for many applications.

At the same time, in spite of the aforementioned progress, the physical reasons for the observed unique properties of polymer nanofibers are poorly understood. Now it is clear that the unusual properties of polymer nanofibers cannot be attributed to the surface effect, and for the explanation of the observed phenomena, the confinement concept seems the most perspective one. Nevertheless, the physical mechanisms providing the confinement effect and, as a result, the appearance of unique properties of polymer nanofibers require thorough investigation.

This book clarifies (maybe, partially) the physical aspects of the problem in question, focusing on the analysis of the size-dependent behavior of electrospun polymer nanofibers.

Arkadii Arinstein

Fall 2017

Chapter 1

Introduction

Polymer nanofibers fabricated with the help of electrospinning have attracted much attention recently because of their unique mechanical and thermodynamic properties compared to the bulk.

The prehistory of the electrospinning process resulting in the fabrication of polymer fibers of sub-micron diameters spans more than 100 years.

For the first time, the process that can be a prototype for electrospinning was mentioned in 1887 when Charles Vernon Boys suggested the process of nanofiber manufacture: He drew the fine quartz filaments by attaching the end of an arrow to a piece of molten quartz and firing it with a crossbow [1]. The first electrospinning patent was filed by John Francis Cooley in the early 1900s [2] (see, for example, Fig. 1.1).

Electrospinning of polymer fibers can be considered a modification or a development of electro-hydrodynamic spraying of liquids when a weakly conducting liquid flowing out of nozzle under high electrical field is sprayed on the very small droplets, which can be precipitated on the opposite electrode, called "collector." In other words, electro-spraying can also be considered a prototype method for electrospinning.

In such a way, in 1914 Eli Franklin Burton and William Bryan Wiegand analyzed the effect of electricity on streams of water drops [3], and in 1914, John Zeleny examined the behavior of fluid droplets

Electrospun Polymer Nanofibers
Arkadii Arinstein

ISBN 978-981-4745-27-7 (Hardcover), 978-1-315-36462-9 (eBook)
www.panstanford.com

at the end of metal capillaries [4]. His effort began the attempt to mathematically model the behavior of fluids under electrostatic forces.

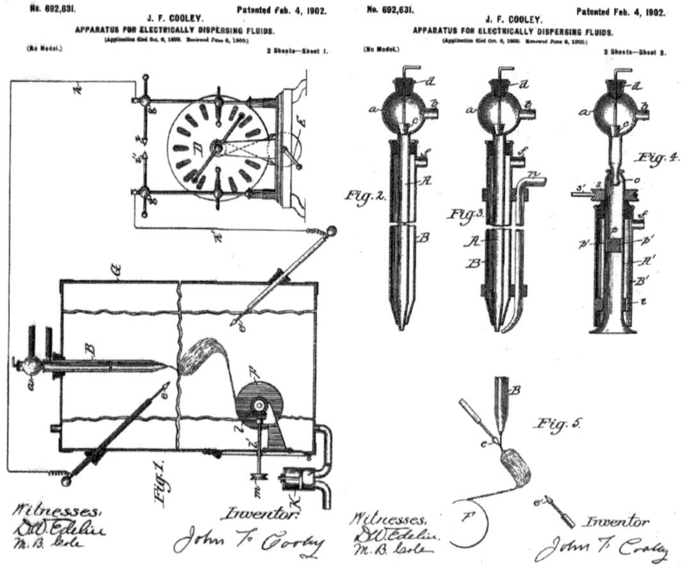

Figure 1.1 US Patent #692631 filed by John Francis Cooley in 1902.

Between 1931 and 1944, Anton Formhals took out at least 22 patents on electrospinning (see, for example, [5]).

In 1938, N. D. Rozenblum and I. V. Petryanov-Sokolov generated electrospun fibers, which they developed into filter materials [6].

Between 1964 and 1969, Sir Geoffrey Ingram Taylor produced the beginnings of a theoretical underpinning of electrospinning by mathematically modeling the shape of the (Taylor) cone formed by the fluid droplet under the effect of an electric field [7].

In the 1990s, several research groups have demonstrated electrospun nanofibers, notably the group of Darrell Reneker who popularized the name "electrospinning" (see, for example, [8]). And only thereafter, the systematic examining of the electrospinning process and of the properties of electrospun polymer nanofibers began. This examining was accompanied by explosive growth of the number of papers devoted to electrospinning and electrospun polymer nanofibers (see Fig. 1.2).

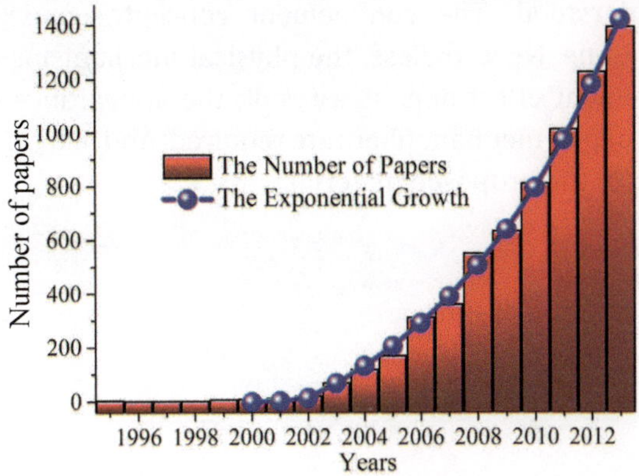

Figure 1.2 Exponential growth of the number of papers devoted to electrospinning and electrospun polymer nanofibers since 2000.

Nowadays, the situation in electrospinning is the following: The hydrodynamics of the electrospinning process is well understandable (in many respects thanks to the experimental studies of Darrell Reneker and, especially, thanks to the theoretical analysis of Alexander Yarin [9]). This fact allows one to fabricate polymer nanofibers in regular and stable way, which stimulated the systematic examination of electrospun fibers. These experimental studies showed that electrospun polymer nanofibers demonstrate outstanding physical properties as compared to bulk specimens corresponding to macroscopic-scale structures. But the most surprising result of the aforementioned studies was the conclusion regarding the effect of the size on the mechanical, thermo-mechanical, and thermodynamic properties of nano-objects; and now the size-dependent behavior of nano-objects is a well-known and widely accepted phenomenon, which was observed in polymer nanofibers [10, 11] and in metallic nanowires [12], as well as in thin polymer films [13]. The most famous example of size-dependent behavior is an abrupt increase in the elastic moduli of different nano-objects when their transversal scale drops below a certain value.

Due to unique properties and possibility of regular and stable fabrication with the help of relative simple equipment, electrospun polymer nanofibers are utilized for many applications. However, in spite of the aforementioned progress, the physical reasons for the observed unique properties of polymer nanofibers are poorly understood. The confinement concept seems the most perspective one. Nevertheless, the physical mechanisms providing the confinement effect and, as a result, the appearance of unique properties of polymer nanofibers are required. And just this question is the focus of the proposed material.

Part I

The Necessary Basics

Chapter 2

Nano-objects

2.1 What "Nano-objects" Mean?

According to the section title, we begin our discussion with the question: What "nano-objects" mean? But why should we ask such a question instead of giving a definition right away and explaining the meaning of the definition.

One might try to give the following definition: "Nano-objects" are those objects that have to be examined in the framework of nanoscience or, more exactly, nanotechnology. But in this case, in place of the initial question, we immediately get a new one: What "nanotechnology" means?

Typing the term "Nanotechnology" in Google search, we get about 37,300,000 results. The feedback for the term "Modern technology" is much impressive: about 635,000,000 results. Comparing these results of our search, we can conclude that nanotechnology amounts almost 6% of all modern technologies. In such a situation, it is clear why the topic of nanotechnology is so attractive for discussions.

Nevertheless, often using the word nanotechnology, many people have only a vague idea about this field of science and technology and can hardly explain what it means. Intuitively, it is clear that nanotechnology deals with very small objects (such small objects were called "nano-objects" earlier). Thus, we come full circle and,

Electrospun Polymer Nanofibers
Arkadii Arinstein

ISBN 978-981-4745-27-7 (Hardcover), 978-1-315-36462-9 (eBook)
www.panstanford.com

perhaps, in such a situation, a definition of nanotechnology should be formulated.

First, according to modern tendencies, let us look in Wikipedia, where we find the following definition (http://en.wikipedia.org/wiki/Nanotechnology):

"Nanotechnology ("nanotech") is the manipulation of matter on an atomic, molecular, and supramolecular scale. ... A more generalized description of nanotechnology was subsequently established by the National Nanotechnology Initiative, which defines nanotechnology as the manipulation of matter with at least one dimension sized from 1 to 100 nanometers."

More exactly, the definition given by the National Nanotechnology Initiative is as follows (http://www.nano.gov/nanotech-101/what/definition):

"Nanotechnology is science, engineering, and technology conducted at the nanoscale, which is about 1 to 100 nanometers."

The website of the European Physical Society (IOP), on the page of the journal Nanotechnology, suggests a definition pointing to the academic aspect of the problem in question (http://iopscience.iop.org/0957-4484):

"Nanotechnology encompasses the understanding of the fundamental physics, chemistry, biology and technology of nanometer-scale objects."

At first glance, such a definition seems nice and clear. Moreover, according to this definition, nano-objects are material samples having sizes of about 1 to 100 nm.

However, all the above definitions focus only on the size of the research/manipulation object, ignoring their intrinsic physical properties. Unfortunately, such an approach can often result in a serious confusion. Indeed, the principal object of our discussion is electrospun polymer nanofibers, and despite the fact that the diameters of these nanofibers can exceed a few hundreds nanometers and even micrometers, nowadays nobody doubts that this research field (electrospinning of polymer nanofibers) should be considered in the framework of nanotechnology.

Before we continue the discussion, let us draw attention to the following well-known physical picture that can be helpful for our goals. Comparing the microworld (the world of extremely small sizes) and our conventional world, we know well that they differ each other not just because of the differences in the sizes of objects: Decreasing the sizes of physical objects up to the atomic scale results in the fact that Newtonian physics governing our conventional (macro) world will be inapplicable and should be replaced by quantum physics. And from the physical point of view, this is the principal difference. This example does not mean that nanotechnology should be reduced to the quantum problem. Nevertheless, it is a good illustration of the fact that a significant decrease in system sizes can result in qualitative alterations in the system behavior, and that physical aspects of the problem in question are of highest importance.

Taking into account all the above discussions, we have to conclude that a more satisfactory and helpful definition of nanoscience and nanotechnology can be achieved only by focusing on the physical aspects of the problem in question, for example, on the intrinsic properties of nanoscale objects. Nevertheless, we can propose the following slightly improved definition: "Nano-objects have at least one dimension of microscopic size." Note that the term "microscopic" should be defined separately, on the base of physical concepts.

Let us consider different types of nano-objects that, in accordance with our definition, can have one, two, or three dimensions of a microscopic scale (see Fig. 2.1).

We start with a bulk sample having the form of a cube with macroscopic sizes in all three dimensions: l_x, l_y, $l_z \propto l_{\text{macro}}$ (see Fig. 2.1a). Reducing the size of this cube in one direction, we get a film with macroscopic sizes in two dimensions— l_x, $l_y \propto l_{\text{macro}}$ — and microscopic size in the third one— $l_z \propto l_{\text{micro}}$. Such a nano-film can be called a two-dimensional (2D) object (see Fig. 2.1b). The next step is further reduction of the size of the obtained film in one more dimension. As a result, we get a nanofiber (or nanowire) having macro-size in only one direction— $l_y \propto l_{\text{macro}}$—and micro-sizes in two others—l_x, $l_z \propto l_{\text{micro}}$. Such objects are often called 1D objects (see Fig. 2.1c). And finally, decrease in the sizes of the initial bulk cube in all directions results in the formation of a nanoparticle, which is of micro-scale in all dimensions: l_x, l_y, $l_z \propto l_{\text{micro}}$. Such nanoparticles can be called 0D objects (see Fig. 2.1d). Note that according to such a classification, a bulk sample has to be called a 3D object.

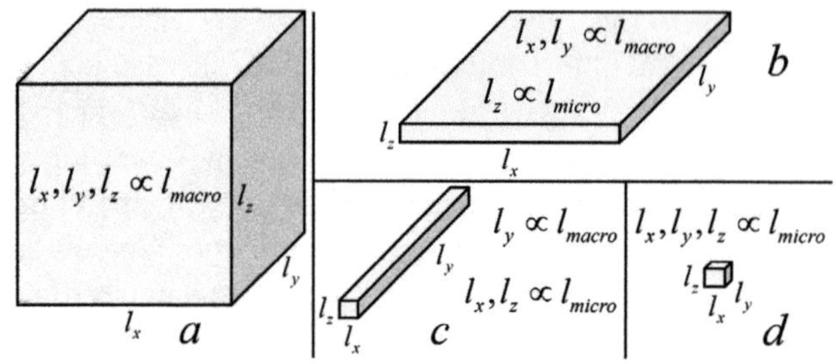

Figure 2.1 Bulk sample of cubic form (a), and different types of nano-objects: nano-film (b), nanofiber or nanowire (c), and nanoparticle (d).

And now, finishing this consideration, we have to define a microscopic size (a micro-scale l_{micro}) introduced earlier for the description of all types of nano-objects.

The first idea is to associate the micro-scale, l_{micro}, with the near-surface layer of the material sample. In such a case, all the material of all nano-objects having at least one dimension of microscopic size is situated in the near-surface layers. Taking into account the well-known fact that the states of atoms and molecules near the surface and inside a sample noticeably differ from each other, the assumption that all the material of nano-objects belongs to the near-surface layers can explain their unique properties. The proposed picture seems quite reasonable in determining nano-objects from the physical point of view.

Nevertheless, the aforementioned definition of the microscopic size (micro-scale, l_{micro}) cannot be accepted for all nano-objects as an universal one. The point is that the near-surface layer has a micro-thickness, $l_{\text{n-s}}$, comparable to the atomic scale (a few angstroms or, as maximum, $1 \div 2$ nm). Therefore, a nano-object consisting only of the near-surface layers should have a thickness less than a few nanometers in at least one direction. Although one can indicate nano-objects that satisfy the aforementioned definition (for example, so-called thin films consisting of a few molecular layers), there are many other objects that are much thicker, but, in accordance with their properties, they should be regarded as nano-objects. For example, we know well that electrospun nanofibers with diameters of a few hundred nanometers and more (even a few micrometers) demonstrate properties that noticeably differ from the bulk samples. This means the introduced restrictions of the possible microscopic sizes are too strong, and we have to modify the definition of the sought-for micro-scale, weakening our requirements.

For example, in place of the requirement that all the material of nano-objects be situated in the near-surface layers, we can assume that the sample surface affects the state of all the material of nano-objects. In other words, the introduced microscopic size is defined by the layers that are affected by the surface of the material. The thickness of such layers can vary from angstroms to hundreds of nanometers. The latter situation (when the thickness of the surface layers reaches hundreds of nanometers) is possible if long-range supermolecular structures are formed inside the sample of the material.

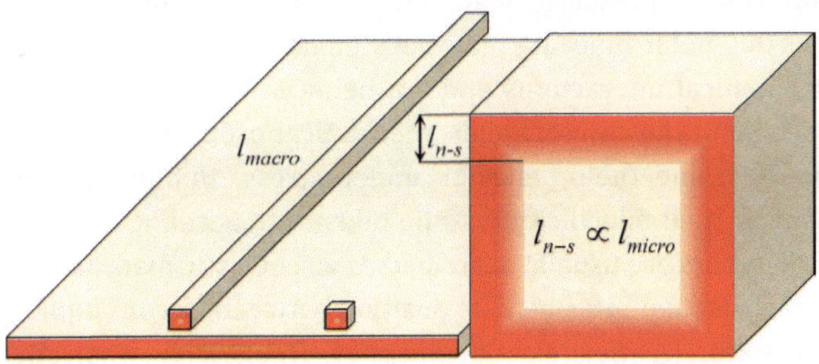

Figure 2.2 The thickness of the layers being affected by the surface of a material.

Utilizing the concept that material layers are affected by the material surface allows one to formulate the following definition of nano-objects based on the physical point of view (see Fig. 2.2): *Nano-objects are those objects all the material of which is affected by their surface.*

2.2 Physical Properties of Nano-objects Compared to the Bulk

In order to better understand what "nano-objects" mean, we will now discuss their intrinsic physical properties and try to explain the features of nanoscale objects from the physical point of view.

First, note that the mechanical and thermodynamic properties of polymer nano-objects regularly show a noticeable modification in comparison to their bulk properties. In the following sections, the mechanical properties of several electrospun nanofibers (as well as of other 1D nano-objects) are reviewed. Next we present experimental results showing the dependence of thermodynamic properties of the nanofibers, such as glass and melting temperature, on the fiber diameter. The related testing methods (both for fiber mats and single nanofibers) will be described in Chapter 3.

2.2.1 Mechanical Properties

Mechanical properties can be defined as the aggregate of indexes that characterize the response of a material under mechanical action (force, pressure, load, etc.). More generally, mechanical properties determine the material's behavior (elastic or inelastic) at mechanical interactions and are being measured in mechanical tests. Like in the case of solids, mechanical properties of polymers also determine their behavior under stress. But in the case of polymers, their specificity is to be taken into account. The point is that polymers are usually described as viscoelastic materials, which emphasizes their intermediate position between viscous liquids and elastic solids. (A linear elastic solid obeys Hooke's law when stress is proportional to strain, whereas a viscous liquid obeys Newton's law when stress is proportional to the rate of change of strain.) Moreover, mechanical properties of polymers depend on their molecular weight, crystallinity, conformational state, etc. At the same time, in this chapter, we discuss only macroscopic properties of polymer materials, ignoring any details on the molecular level.

Thus, mechanical properties of polymers depend on their state. This fact (a polymer state affects its properties) is of high importance as in reality, almost all polymers demonstrate a nonequilibrium state that, being frozen below glass transition temperature, is determined by their pre-history. Indeed, in such a situation, the polymer properties cannot be determined only in the thermodynamic equilibrium; they should be determined in every, generally, nonequilibrium physical state of a system under consideration. Nevertheless, in the thermodynamic equilibrium, the polymers demonstrate their intrinsic properties, whereas the measured data for specific system can vary. These variations can reflect the peculiarities of the state of the polymer system, and such an interpretation seems reasonable from the physical point of view.

Assuming low deformations of tested polymer, one can introduce the elastic modulus (Young's modulus) of a polymer material (in both equilibrium and nonequilibrium). In the case of high deformations, additional characteristics should be introduced, such as, yield strength, tensile strength, etc. The last situation (high deformations) corresponds to non-linear behavior of polymers, whereas in the case of small deformations, polymers demonstrate the simplest linear behavior, and this simplest case is suitable for our analysis. In addition, note that in the case of small deformations, the elasticity of polymers is dominant, whereas with an increase in deformation degree, the polymers start to demonstrate their plasticity, i.e., viscous properties start to be dominant.

Thus, we analyze the behavior of the elastic modulus (Young's modulus) of nanofibers, and this analysis shows the unique mechanical properties of nanofibers compared to bulk.

One of the first measurements of the tensile elastic modulus of fiber-like nano-objects was presented by Cuenot et al. in 2000 [14]. In this examination, polypyrrole nanotubes were mechanically tested in three points bending, using atomic force microscopy (see Fig. 2.3).

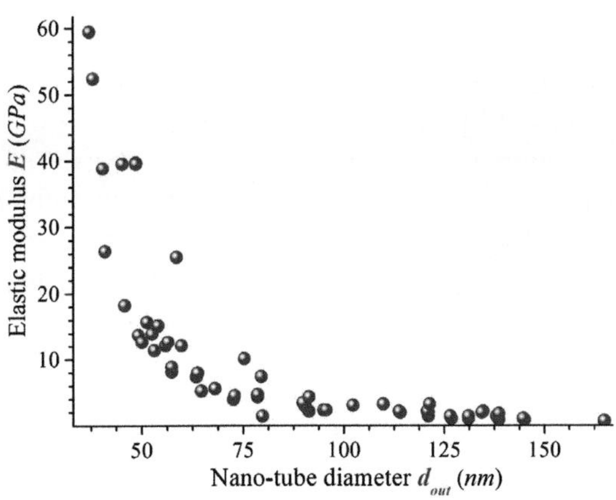

Figure 2.3 Elastic modulus, *E*, of polypyrrole nanotubes, deduced from the force curve measurements as a function of their outer diameter d_{out}. Reprinted from Ref. [14], Copyright 2000, with permission by the American Physical Society (https://doi.org/10.1103/PhysRevLett.85.1690).

More detailed analysis of the observed phenomenon was presented in the next publication of this group where the elastic modulus for Ag nanowires as a function of their diameter was examined (see Cuenot et al. [11]). The variations in the measured moduli have demonstrated the same regularity as in the case of polypyrrole nanotubes: The elastic modulus of Ag nanowires increased sharply when their diameters became small enough (see Fig. 2.4).

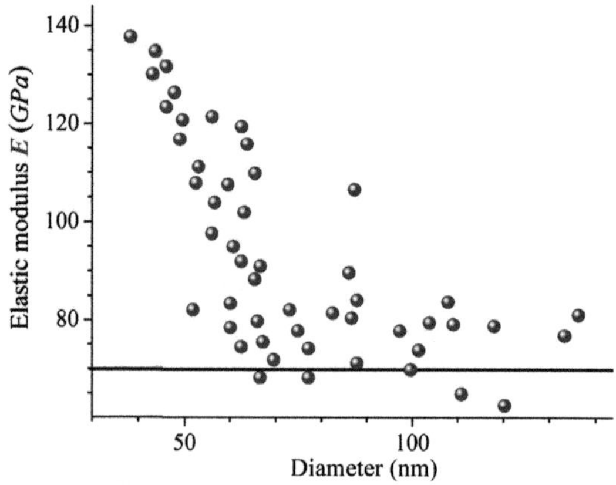

Figure 2.4 Variation of the measured elastic modulus for *Ag* nanowires as a function of their diameter. Reprinted from Ref. [11], Copyright 2004, with permission by the American Physical Society (https://doi.org/10.1103/PhysRevB.69.165410).

Now we will not analyze the explanation of the observed behavior, proposed by Cuenot et al. in [11], when the increase in the elastic modulus is attributed to surface effects. The impact of surface tension on the properties of fiber-like nano-objects as well as the physical reasons of the phenomenon in question will be discussed in details in the Part III, Section 6.2 "Do surface effects explain the unique elasticity of polymer nanofibers?" and Section 7.1 "Confinement mechanism of electrospun polymer nanofibers reinforcement."

In spite of the fact that no electrospun polymer nanofibers were examined in the aforementioned studies (polypyrrole nanotubes and Ag nanowires were tested), these results are of high interest for us, demonstrating some general properties of nano-objects.

After a few years, this unusual phenomenon was accepted by the scientific community and gave rise to further studies on nanofiber properties.

For example, in 2005, Tan and Lim examined variation in the elastic modulus of poly-L-lactic acid (PLLA) nanofibers with fiber diameter, using the nanoindentation technique [15] (see Fig. 2.5).

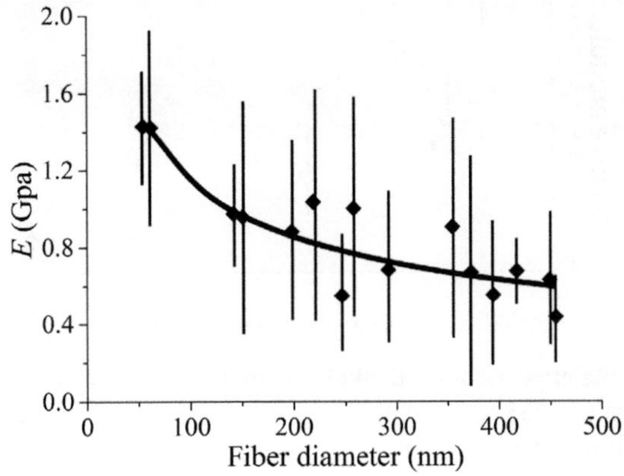

Figure 2.5 Variation of elastic modulus with fiber diameter for nanoindentation of PLLA nanofibers. Reprinted from Ref. [15], with the permission of AIP Publishing.

In 2006, Shin et al. studied the elastic modulus of nanofibers of polyAMPS acid, with the help of the three-point bending test [16] (see Fig. 2.6).

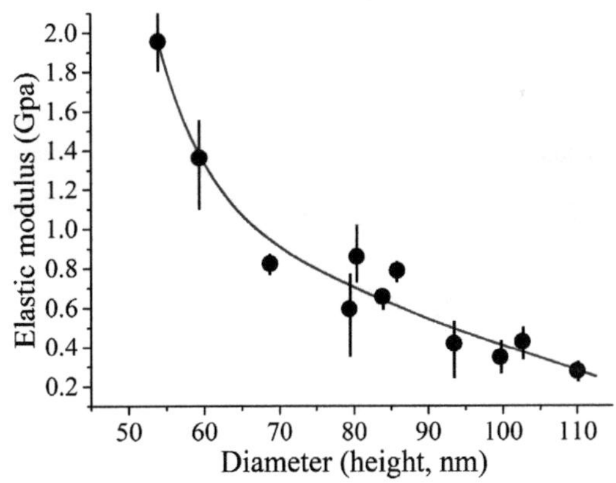

Figure 2.6 Three-point bending test of polyAMPS acid. Reprinted from Ref. [16], with the permission of AIP Publishing.

In the same year, Ji et al. demonstrated the relative modulus G of electrospun PS fibers as a function of the fiber diameter [17] (see Fig. 2.7).

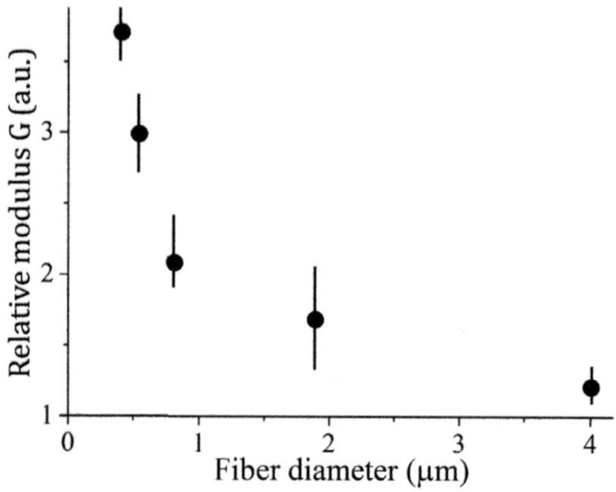

Figure 2.7 Relative modulus G of electrospun PS fibers as a function of the fiber diameter at 295 K. Reprinted with permission from Ref. [17]. Copyright 2006, American Chemical Society.

One year later, Arinstein et al. also demonstrated a similar dependence of Young's modulus E versus diameter d of electrospun Nylon-6,6 nanofibers [12] (see Fig. 2.8).

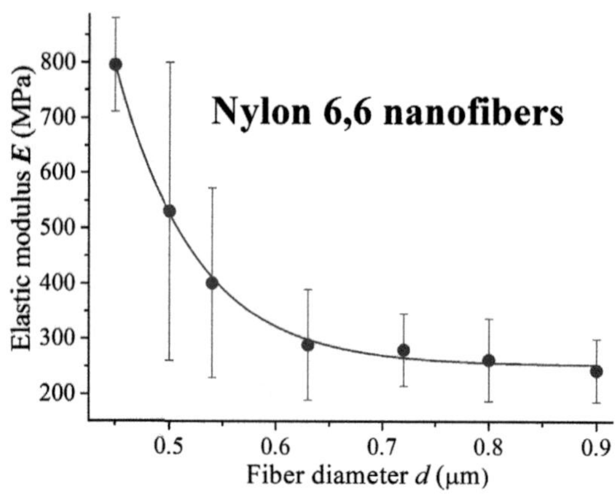

Figure 2.8 Young's modulus E versus diameter d of electrospun Nylon-6,6 nanofibers. Error bars show the standard deviation. Reprinted from Ref. [12], Copyright 2007, Nature Publishing Group.

A similar behavior was also discovered by other authors (see, for example, [18–22]). All these observations have demonstrated the same regularity: Electrospun polymer nanofibers exhibit enhanced mechanical properties compared to those of their macroscopic counterparts.

2.2.2 Thermodynamic Properties

Certain thermodynamic properties of nanofibers also demonstrate unusual behavior.

For example, the temperature dependences of elastic modulus have been presented by Ji et al. in [17], where it has been shown that these dependences differ for fibers of small versus large diameters at temperatures below their glass transition temperature, T_g. The elastic modulus of fibers of 4 μm diameter increases with elevating temperatures, corresponding to rubber elasticity, whereas it decreases for fiber diameters <1 μm, a behavior common to solid and crystalline materials (see Fig. 2.9).

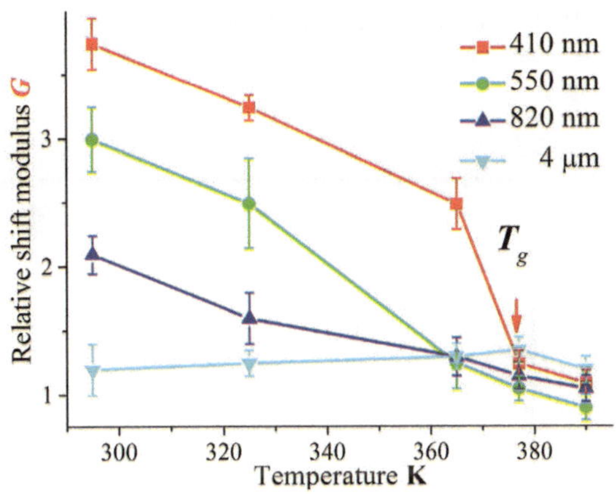

Figure 2.9 Relative shear modulus *G* of electrospun polystyrene fibers as a function of the temperature, at different fiber diameters. Reprinted with permission from Ref. [17]. Copyright 2006, American Chemical Society.

The same data, presented in another form, demonstrate that modulus dependences on fiber diameter are modified as temperatures increase: At room temperature, elastic modulus monotonically increases as diameters decrease, and at temperatures above T_g, the dependence becomes non-monotonic (see Fig. 2.10).

Besides, a shift in the phase transition temperatures in electrospun polymer nanofibers can be observed. So the depression of melting temperature as a function of the diameter of electrospun semi-crystalline electrospun polyethylene oxide (PEO) nanofibers was demonstrated by Wang and Barber [23] (see Fig. 2.11).

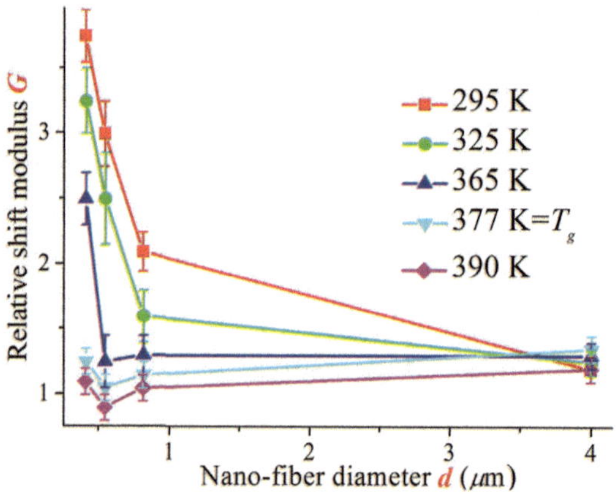

Figure 2.10 Relative shear modulus *G* of electrospun polystyrene fibers as a function of the fiber diameter, at different temperatures. Based on the data presented in Fig. 2.9.

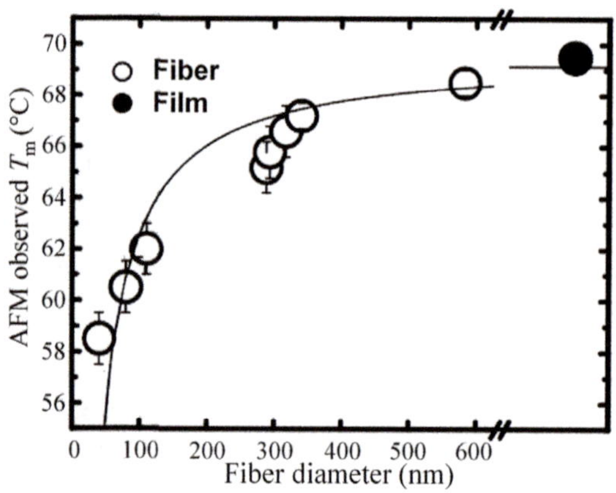

Figure 2.11 Plot of T_m as a function of fiber diameter for electrospun PEO fibers. The curve represents a least-squares fit of the data to Eq. (7.32), resulting in estimating parameter values of $S_{vib}(\infty) = 19.27\ J \cdot mol^{-1} \cdot K^{-1}$ and $D_0 = 1.68$ nm (see Section 7.2 "Shifting of the melting point for semi-crystalline polymer nanofibers"). Reprinted with permission from Ref. [23], Copyright 1990, CCC Republication.

The glass transition temperature, T_g, in polymer nanofibers also differs compared to bulk. So using atomic force microscopy (AFM), Wang and Barber have shown that the glass transition temperature, T_g, of electrospun polyvinyl alcohol (PVA) fibers starts to decrease for smaller fiber diameters, especially for fiber diameters below 250 nm [24] (see Fig. 2.12).

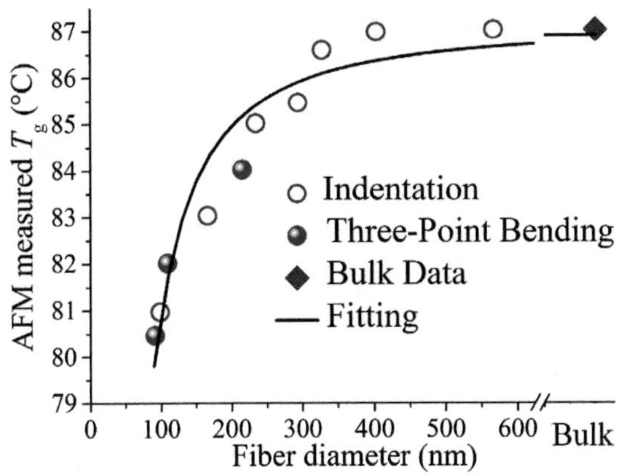

Figure 2.12 Plot of variations in glass transition temperature, measured using AFM indentation and three-point bending testing, with electrospun PVA fiber diameter. Reprinted from Ref. [24], Copyright 2011, with permission from John Wiley and Sons.

2.2.3 Size-Dependent Behavior of Nano-objects

Due to the fact that the objects of our interest are polymer nanofibers, all the aforementioned examples of unusual behavior of nano-objects are related to nanofibers or fiber-like objects. Nevertheless, other types of nano-objects (thin films and nanoparticles) also demonstrate similar unusual behavior: Thermo-mechanical properties of such nano-objects differ compared to their bulk counterparts (the corresponding literature can be found without any problems). Therefore, our conclusions are related to all types of nano-objects: thin films, nanofibers, and nanoparticles.

The first conclusion on the basis of all the aforementioned experimental observations is that the properties of nano-objects sharply vary when their scale drops below a critical threshold (e.g., thickness for films and diameters for fibers or nanoparticles). In spite of such a behavior, we do not deal with critical phenomena; rather, a crossover occurring in a specific range is observed. In addition, below crossover, the variations in nano-object properties depend on the micro-scale, l_{micro}, of the nano-object: They increase with a decrease in this micro-scale.

Such a feature of nano-objects is called size-dependent behavior and gives rise to numerous discussions. In spite of these discussions, the physical reasons for the observed size-dependent changes in the nano-object properties are poorly understood. Mechanistic models have attributed such behavior to surface effects for both 2D and 1D nano-systems. While surface phenomena clearly play a role in the formation of the properties of materials of reduced sizes, detailed analysis shows that surface effect alone cannot adequately explain the experimental observations. Therefore, additional ideas should be utilized in order to understand the observed size-dependent behavior. One of such ideas is the confinement concept, which explains the size-dependent behavior by geometrical restrictions.

Chapter 3

Experimental Examination of Electrospun Polymer Nanofibers

Each examination of electrospun nanofibers requires testing of these fibers by different experimental methods. The first information about the properties of our object can be obtained with the help of imaging, including optical microscopy (see, for example, Figs. 4.5 and 5.14), scanning electron microscopy (SEM) (see Fig. 3.1, as well as Figs. 4.21, 4.25, 4.30b, 5.1, 5.13, and 5.15), and transmission electron microscopy (TEM) (see Fig. 3.2).

In SEM, a type of electron microscope produces images of a sample by scanning it with a focused beam of electrons. SEM images contain information about the morphology of tested samples (see Figs. 4.21 and 5.13), as well as about the topography of their surface (see, for example, Figs. 3 and 4 in [26]) or of their cross sections (see Figs. 4.25, 4.30b, and 5.1).

TEM is a microscopy technique in which a beam of electrons is transmitted through an ultra-thin specimen, interacting with the specimen as it passes through it. TEM images allow one to analyze fine details of internal structure of electrospun nanofibers: different types of supramolecular structures, spatial distribution of different impurities, nanoparticles, carbon nanotubes, etc. (see Fig. 3.2).

Electrospun Polymer Nanofibers
Arkadii Arinstein

ISBN 978-981-4745-27-7 (Hardcover), 978-1-315-36462-9 (eBook)
www.panstanford.com

Nevertheless, in order to study mechanical and thermo-mechanical behavior of any physical object (in our case, electrospun polymer nanofibers) and also to describe their structural features, quantitative methods of analysis are required. The modern instrumentation arsenal for studies of different polymers offers a multitude of possibilities for examining electrospun fibers: Dynamic mechanical analysis (DMA), differential scanning calorimetry (DSC), small and wide angle X-ray scattering (SAXS and WAXS), different types of spectroscopies, such as infrared (IR) and ultraviolet (UV) spectroscopies, polarized Raman spectroscopy, and nuclear magnetic resonance (NMR) are applicable for our goals.

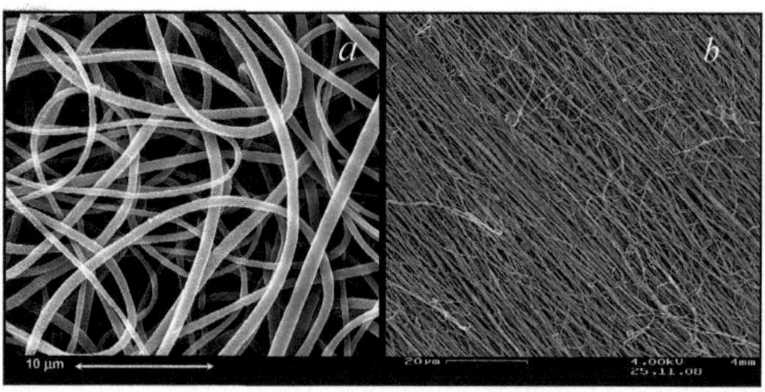

Figure 3.1 SEM images of electrospun polymer nanofibers: randomly collected (a) and aligned mats (b).

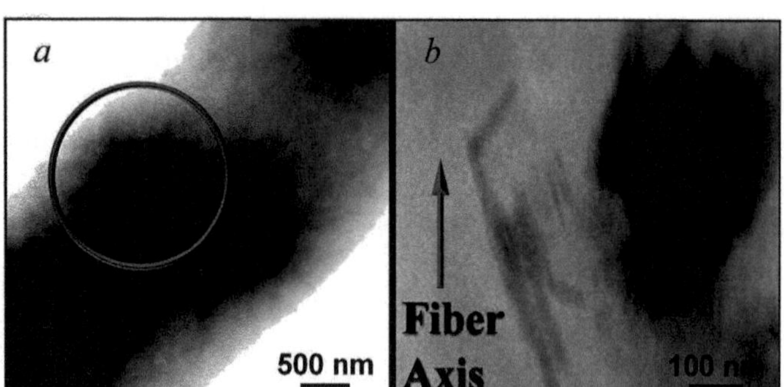

Figure 3.2 TEM images of the electrospun 7.5 wt % PEVA260 fibers (a) without and with (b) 6.6 wt % clay. Reprinted from Ref. [25], Copyright 2001, with permission from John Wiley and Sons.

3.1 Experimental Examination of Nanofiber Mats

It seems that the simplest way to examine mechanical and thermo-mechanical properties of electrospun nanofibers is to collect enough amount of material in the form of a fiber mat (see Fig. 3.1), and thereafter one can utilize this mat, like a film, for necessary measurements. For example, thermo-mechanical behavior of electrospun fiber mats can be studied with the help of DMA.

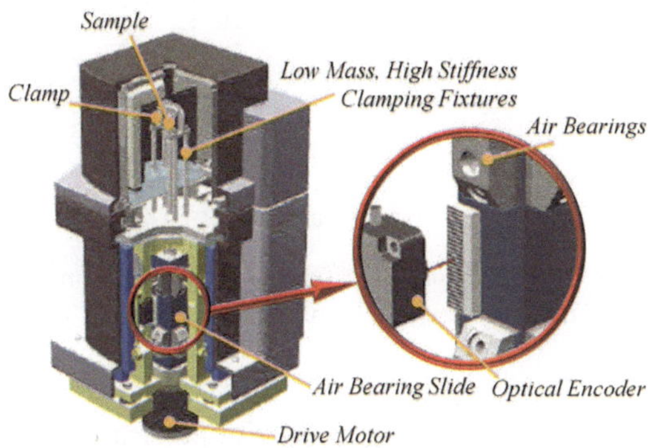

Figure 3.3 The view of central module of a DMA device.

A DMA device allows one to measure the storage (E') and loss (E'') moduli of specimens at different temperatures as well as to determine the temperatures of possible phase transitions: the glass transition temperature, the melting point, the crystallization temperature, etc. (see Fig. 3.4).

However, the situation is not so simple as it seems. The point is that the mat structure can significantly affect the measurement results. Indeed, a deformation (stretching) of a randomly collected fiber mat (see Fig. 3.1a) can result in the straightening of curved fibers; the slipping of the fibers relative to each other in some linking points is also possible. The situation can dramatically change if we try to test a mat consisting of well-aligned oriented electrospun nanofibers (see Fig. 3.1b). In this case, a small stretching of such a mat results in no modification of its structure; therefore, only a small tensile deformation of parallelly aligned fibers is possible. Thus, DMA test allows one to analyze successfully a mat consisting of well-aligned oriented electrospun nanofibers. For example, the elastic modulus of the fibers forming the tested sample can be measured on the basis of uniaxial stretching. In doing so, only the porosity of the tested mat is to be taken into account.

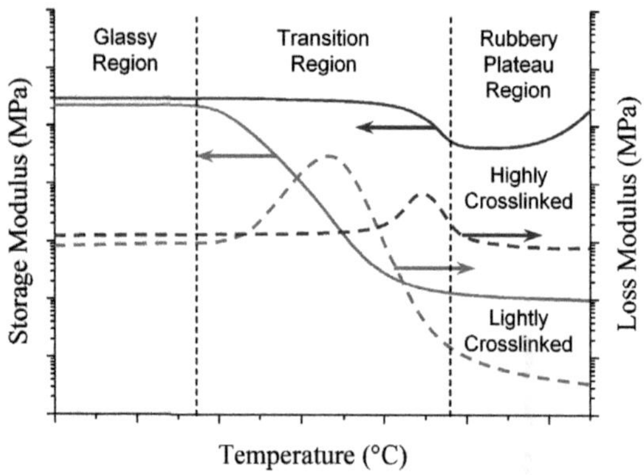

Figure 3.4 The typical results of a temperature sweep in DMA test for a polymer sample. Storage modulus (E') and loss modulus (E'') are plotted against temperature.

Note that there are a few methods to fabricate mats with well-aligned oriented electrospun nanofibers. For this goal, rotating collectors, such as rotating cylinders, wheels, etc., are usually utilized (see Fig. 3.5).

Thus, the DMA test allows one to analyze a mat consisting of well-aligned oriented electrospun nanofibers, measuring, for example, the fiber's elastic modulus on the basis of uniaxial stretching. At the same time, determination of the temperatures of the phase transitions in the material of the electrospun fibers does not requires the fabrication of aligned mat: randomly collected mats are suitable for this goal.

The phase transitions in the material of the electrospun fibers can also be examined with the help of DSC (see Fig. 3.6); in doing so, randomly collected mats are also suitable.

Randomly collected mats allow analysis of the crystalline structure of the fiber polymer matrix with the help of WAXS and SAXS (crystallinity level, some crystallite parameters, etc.), whereas the examination of orientational effects requires well-aligned fiber mats.

Figure 3.5 The device for the fabrication of well-aligned fiber mats where the rotating wheel is utilized as the collector. Additional advantage of such a collector is that the electric field is concentrated on the wheel sharp edge, promoting a better fiber alignment.

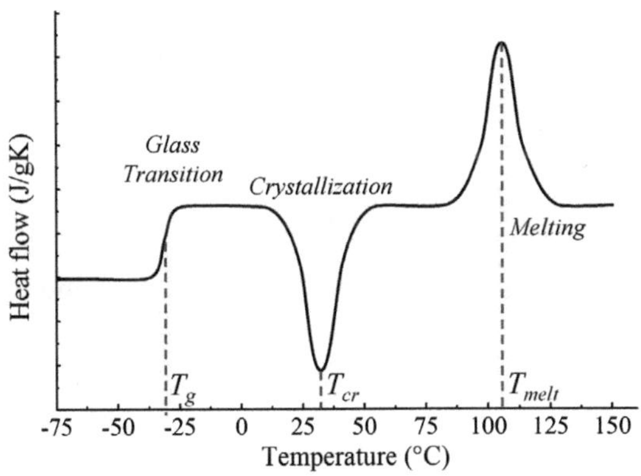

Figure 3.6 A schematic DSC curve using the baseline as the reference. Three types of phase transitions are presented: glass transition, crystallization, and melting.

Thermo-mechanical testing of randomly collected mats can be realized with the help of shear modulation force microscopy (SMFM) elaborated by Ge et al. [27] on the basis of atomic force microscopy (AFM). The experimental setup of the SMFM method is pictured in Fig. 3.7. A sinusoidal electrical signal applied to the piezocrystal, which controls the X-displacement of the cantilever, results in a small oscillatory motion of the tip parallel to the sample surface. A normal load force maintains good contact of the tip with the sample. The X-amplitude response being measured by the photodiode detector allows one to calculate the necessary parameters of the sample material.

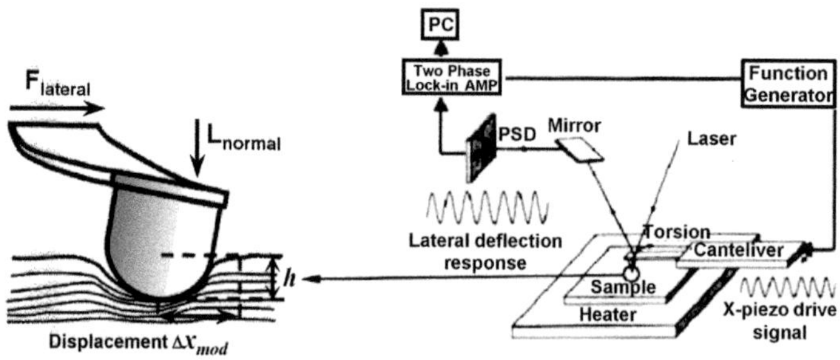

Figure 3.7 Principle of shear modulation force microscopy. Reprinted with permission from Ref. [17]. Copyright 2006, American Chemical Society.

3.2 Experimental Examination of Single Nanofiber

3.2.1 Atomic Force Microscopy

The elastic modulus of single electrospun nanofiber can be measured with three-point bending using AFM. For the first time, such a measurement method was demonstrated by Cuenot et al. by the mechanical testing of polypyrrole nanotubes, suspended over pores of a substrate [14].

The principal scheme of the AFM three-point bending test is presented in Fig. 3.8.

Below, we discuss one of the possible interpretations of the AFM measurement proposed by Cuenot et al. [14]. Considering a testing nano-cylinder (nanofiber or nanotube) as a beam, we can adopt the well-known mechanical model to obtain the tensile modulus E from the measured force F and deflection δ. Assuming a force F applied at the beam midpoint and inducing a deflection δ, an expression for the elastic energy, U_{el}, of the bent beam is

$$U_{\rm el} = \frac{1}{2}k\delta^2 - F\delta, \tag{3.1}$$

where k is the beam elastic stiffness. Due to the good adhesion of the beam to the substrate, the boundary conditions correspond to the clamped-beam model (otherwise, when adhesion is too weak, the beam can slip, and no testing is possible). In this case, the beam elastic stiffness is

$$k = \frac{192EI}{L^3}, \tag{3.2}$$

where $I = \pi D^4/64$ is the moment of inertia of the section, L and D are the length and diameter of the beam, respectively.

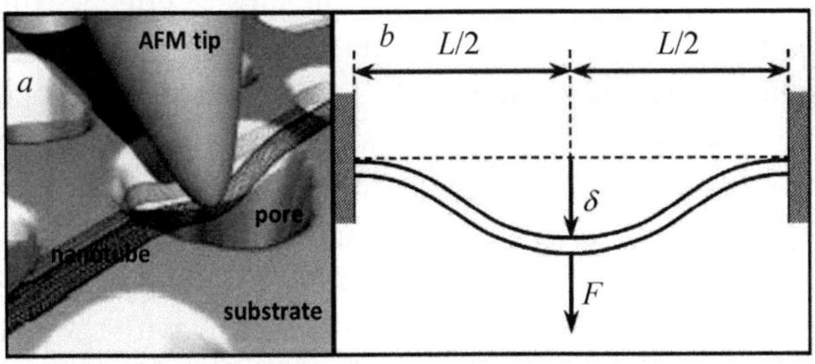

Figure 3.8 (a) Schematic representation of the AFM three-point bending test. (b) Scheme of beam deformation in the three-point bending test.

As deformation of the beam induces an increase in its area, surface tension should be taken into account (let us remember that in the case of nano-objects, the surface effects can be noticeable). Therefore, the elastic energy U_{el} (3.1) should be modified, and the total energy U of the bent beam is

$$U = \frac{1}{2}k\delta^2 - F\delta + \gamma\Phi\Delta L(1-\nu)$$

$$= \frac{3\pi D^4}{2L^3}\left[E + \frac{8\gamma(1-\nu)L}{5D^3}\right]\delta^2 - F\delta, \tag{3.3}$$

where $\Phi = \pi D$ is the perimeter of the beam cross section, and $\Delta L = 12\delta^2/5L$ is the beam elongation.

The first derivative of Eq. (3.3) results in the linear relationship between the measured force F and deflection δ:

$$F = \frac{3\pi D^4}{L^3}\left[E + \frac{8\gamma(1-\nu)L}{5D^3}\right]\delta \tag{3.4}$$

Thus, we can conclude that due to the surface tension effect, the elastic modulus of the nano-beam, measured with the AFM three-point bending test, depends on its diameter:

$$E_{\text{eff}} = E + \frac{8\gamma(1-\nu)L}{5D^3} \tag{3.5}$$

Equation (3.5) seems to explain the aforementioned size-dependent behavior of nano-objects in mechanical tests. Nevertheless, we can indicate some experimental data that contradict to this conclusion, demonstrating its fallibility (this question will be discussed in detail in Section 6.2 "Do surface effects explain the unique elasticity of polymer nanofibers?").

Finishing the description of AFM applied to electrospun nanofiber testing, it is suitable to mention the indentation method used for nanomechanical thermal analysis of electrospun polymer fibers by Wang et al. [28]. In this paper, the thermo-mechanical response of individual fibers was examined by locating an individual electrospun fiber by AFM imaging and indenting with the AFM tip.

In the indentation experiment, the AFM tip indentation depth δ is produced by a corresponding force F consisting of the externally applied force F_e and the adhesion force F_a. The first force, F_e, quadratically depends on the difference $\delta - \delta_r$, whereas the second one, F_a, depends on this parameter linearly (for details, see [28]):

$$\begin{aligned} F(\delta) &= F_e + F_a \qquad (3.6) \\ &= \frac{2E\tan\alpha}{\pi(1-\nu^2)}(\delta-\delta_r)^2 - \frac{8\gamma_a\tan\alpha}{\pi\cos\alpha}(\delta-\delta_r) \end{aligned}$$

where E is the elastic modulus of the fiber, α is the half tip cone angle, ν is Poisson's ratio of the fiber, γ_a is the thermodynamic work of adhesion, and δ_r is the AFM tip depth at $F = 0$ on the unloading curve.

Figure 3.9 shows a typical force–displacement curve produced when pushing the AFM tip into the PEO fiber and the subsequent removal of the tip at 40°C. The loading curve shows how an increase in force is required to push the AFM tip into the fiber with the maximum indentation depth δ_m achieved at a maximum applied force F_m. The unloading curve differs from the loading curve due to the viscoelastic response of the fiber.

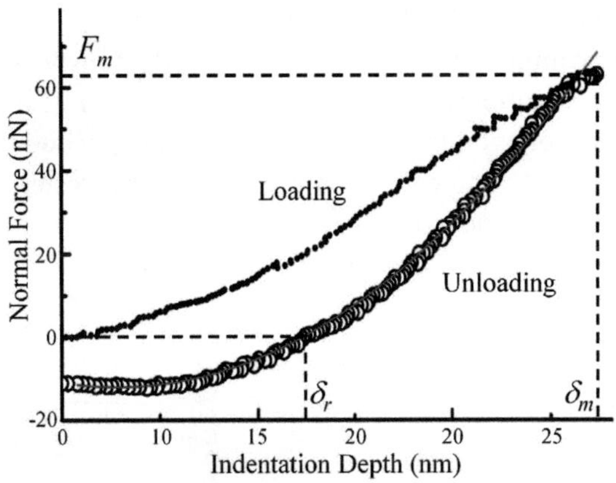

Figure 3.9 Plot of the loading–unloading normal force-indentation depth $F - \delta$ for the indentation of an electrospun PEO fiber at 40°C showing the fitting curve with the theoretical $F - \delta$ dependence of Eq. (3.7). δ_m is the maximum indentation depth observed by the maximum applied force F_m. Reprinted from Ref. [28], with the permission of AIP Publishing.

The AFM indentation method can be applied also for testing cast films and electrospun fiber mats. From this point of view, the AFM indentation method is similar to SMFM discussed in Section 3.1 "Experimental examining of nanofiber mats."

3.2.2 Tensile Test

In order to exclude the surface tension effect, a tensile deformation should be adopted for the mechanical testing of nanofibers. The point is that in the tensile deformation mode, the additional surface energy U_{sf} depends on the elongation strain ΔL, only linearly:

$$U_{sf} = \frac{4\gamma}{D}\frac{\Delta L}{L}, \tag{3.7}$$

and does not contribute to effective elastic modulus being calculated with the help of the second derivative of the total energy U_{tot}, which includes both the elastic energy U_{el} and the additional surface energy U_{sf}:

$$U_{tot} = U_{el} + U_{sf} = \frac{E}{2}\left(\frac{\Delta L}{L}\right)^2 + \frac{4\gamma}{D}\frac{\Delta L}{L} \quad (3.8)$$

(this point will be discussed in Section 6.2 "Do surface effects explain the unique elasticity of polymer nanofibers?").

In spite of multitude technical problems, the tensile test of single electrospun nanofibers was executed by Zussman et al. [29]. The stress–strain dependencies, obtained in such experiments (see Fig. 3.10), allow to determine the elastic modulus of the tested nanofibers according to the slope of the plotted curves.

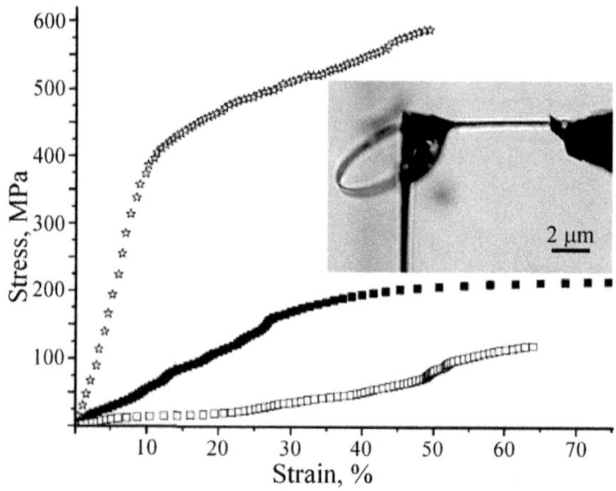

Figure 3.10 Typical stress–strain curve of the individual PA66 nanofibers collected at take-up velocities of 5 (□) and 20 (■) m/s, and a commercial microfiber prepared by melt extrusion processing followed by cold drawing (*). The inset demonstrates image taken during a tensile test. The fiber was attached to the tip of an AFM cantilever and pulled transversally by a stainless steel wire attached to the other end of the fiber. Reprinted from Ref. [29], Copyright 2006, with permission from John Wiley and Sons.

3.2.3 Resonance method

Thus, there are no doubts that tensile analysis [12, 19, 29–31] is the most appropriate method to study the mechanical properties of individual polymer nanofibers. Unfortunately, tensile tests often result in the destruction of the specimen, whereas vibration-based methods allow one to perform non-destructive tests.

A vibration method utilizing two micro-cantilevers connected by a nanofiber was suggested by Yuya et al. [32]; however, realization of this method seems problematic due to the suggested mounting design.

An additional advantage of vibration methods is the possibility to utilize the resonance effect, which noticeably increases the measurement accuracy. When searching for a new measurement method, it is suitable to direct our attention to the uncommon features of a string pendulum, which can increase (by orders of magnitude) the resolution of traditional resonant vibration techniques [33]. This feature of the string pendulum appeared very helpful for the elaboration of novel resonant method for the tensile testing of a single nanofiber [34].

The proposed experimental system consists of an individual nanofiber with a ball attached to its end. The other end of the nanofiber is attached to a rigid cantilever beam located on a piezoelectric-actuated base (see Fig. 3.11a). Such a system can be regarded as a string pendulum consisting of a string of length l_0 and stiffness constant κ, with a suspended bob of mass m, where the string elongation Δl at equilibrium is $\Delta l = mg/\kappa$ (see Fig. 3.11b). Forced longitudinal oscillations of a string pendulum can contain two modes of motion: a regular oscillation under the string's elastic force, and free flight only under gravity. The regular oscillations occur when the oscillation amplitude is small and the nanofiber is still stretched (see Fig. 3.11c). In the event of large oscillation amplitudes that exceed the string elongation caused by gravity, the bob can reach a height at which the elastic force no longer affects it, and any further motion occurs only under gravity, i.e., the bob executes free flight (see Fig. 3.11d).

When large oscillation amplitudes occur, the flight height of the suspended mass considerably exceeds the amplitude of regular oscillations. Our interest is focused solely on this effect. The ratio between flight height h and oscillation amplitude Δx (the amplification coefficient) can be estimated using the law of energy conservation. The elastic energy $\kappa\Delta x^2/2$ is transformed into potential energy of the bob, mgh, so the amplification coefficient is equal to the ratio of the oscillation amplitude to the string elongation:

$$\frac{h}{\Delta x} \sim \frac{\kappa \Delta x}{mg} = \frac{\Delta x}{\Delta l}, \tag{3.9}$$

which, in the case of high rigidity of the string, can amount to 10–100.

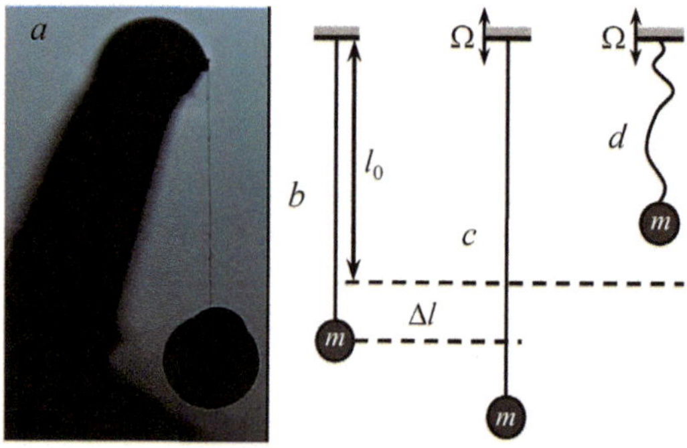

Figure 3.11 The string pendulum. (a) View of a ball glued to a nanofiber and suspended from a cantilever beam, which is attached to the piezoelectric-actuated base (×100). (b) The equilibrium state of the pendulum with string elongation Δl. (c) The downward displacement of the bob, at whose length the string remains stretched. (d) The upper displacement of the bob's flight when the string becomes unstretched. Reprinted from Ref. [34], with the permission of AIP Publishing.

For modeling the system, we locate the origin of the coordinate system at the suspension point with the upward x-axis. The equilibrium position of the bob is defined by $x_0 = -(l_0 + \Delta l)$. The displacement of the oscillating bob is Δx, so its coordinate is $x = x_0 + \Delta x = \Delta x - (l_0 + \Delta l)$. If the suspension point undergoes oscillations $A \sin(\Omega t)$, the resultant force acting on the bob is $F = \kappa [A \sin(\Omega t) - \Delta x]$ and the motion equation of this system when the string is stretched is

$$m\frac{d^2 \Delta x}{dt^2} = \kappa\left[A\sin(\Omega t) - \Delta x\right] - \mu\frac{d\Delta x}{dt} \tag{3.10}$$

where μ is the effective dissipation coefficient.

When the string is unstretched, the bob moves only under gravity:

$$m\frac{d^2 \Delta x}{dt^2} = -mg \tag{3.11}$$

Using a dimensionless displacement $u = \Delta x / \Delta l = \Delta x \omega_0^2 / g$, and dimensionless time $\tau = \omega_0 t$ ($\omega_0^2 = \kappa / m$ is the natural frequency of the oscillating bob), Eqs. (3.10) and (3.11) can be rewritten as:

$$\frac{d^2u}{d\tau^2} + 2\gamma \frac{du}{d\tau} + u = a\sin(\Omega_0 \tau), \quad u < 1 + a\sin(\Omega_0 \tau), \tag{3.12}$$

$$\frac{d^2u}{d\tau^2} = -1, \quad u > 1 + a\sin(\Omega_0 \tau), \tag{3.13}$$

where $\gamma = \mu/2m\omega_0$ is the damping factor, $\Omega_0 = \Omega/\omega_0$, and $a = A / \Delta l = A\omega_0^2 / g$.

Note that the modes of oscillation and free flight alternate with each Other. Thus, the initial conditions for each stage are determined by the motion of the previous one. In doing so, due to the nonlinear boundary conditions, the presented system is highly nonlinear. Details of the theoretical analysis of the system are presented in [33].

The fact that, due to its flights, the displacement of the oscillating bob noticeably exceeds the amplitude of regular forced oscillations, and since this motion is of a resonance type, allows one to ascertain the natural resonance frequency of the string. As a result, Young's modulus of the fiber material can be determined.

The mean flight heights $\bar{h}$, which was obtained assuming that the damping factor is small ($\gamma << 1$), the frequency Ω_0 is close to the resonant ($|1 - \Omega_0^2| \lesssim \gamma^2$), and the excitation amplitude A is small in comparison to the flight heights of the bob, is the following [33]:

$$\bar{h} = \frac{a^2 \Omega_0^2 \sin^2(\pi \Omega_0 / 2)}{\pi\gamma \left[(1 - \Omega_0^2)^2 + (2\gamma\Omega_0)^2 \right]} \tag{3.14}$$

The above dependence is of a resonance type with dimensionless resonance frequency $\Omega_0^{(res)} = \Omega_{res}/\omega_0 = 1$, where Ω_{res} is the measured resonance frequency of the system. Thus, the measurement of $\bar{h}$ allows one to determine the natural frequency ω_0 of the string pendulum.

The elastic modulus of the nanofiber can be calculated from the following equation:

$$E_f = \Omega_{res}^2 m l_0 / A_f \tag{3.15}$$

where A_f is the nanofiber cross-sectional area.

The experimental system having demonstrated the efficiency of the proposed method [34] consisted of a 3 DOF (Newport) precision manipulator, which was used for handling and attaching the measured nanofiber to a ball (Au-Sn) on one side and to a rigid cantilever beam (stainless steel) on the other side. The cantilever was mounted on a piezoelectric-actuated base plate (see Fig. 3.11a). The nanofiber was electrospun from a Nylon-6,6 solution (see Ref. [60]). The experimental setup was assembled under an optical microscope (Olympus BX50). The measurement procedure was accomplished with an attached CCD camera (DP71). A wave generator (Agilent 33220A) and signal amplifier (A.A. Lab Systems) were used to actuate the piezoelectric base. The measurement process was fully automated and controlled by software developed under C++, Visual Studio platform.

The bob's mass and the length of the nanofiber were selected so that the effective system resonance along the nanofiber's axis was relatively distant from the transverse string oscillations, the planar pendulum oscillations, and the resonances of other components of the experimental system.

The piezoelectric base was actuated by a sine signal of a given amplitude, and forward and backward scanning of a defined interval of frequencies was performed. For each scanned frequency, the averaging still image (exposure time $\sim$10 sec.) was taken (see Fig. 3.12). This process was repeated for different actuation amplitudes to find the lowest amplitude where the free-flights of the ball were still recognized in order to minimize side effects and to obtain the most regular system behavior.

Typical plots of ball flight heights versus input signal frequency $\nu = \Omega/2\pi$ are presented in Fig. 3.13. The position of the peak corresponds to the natural frequency of the tested nanofiber. The obtained resonance frequency allows one to find Young's modulus of the nanofiber using Eq. (3.15). This procedure was performed for nanofibers with different diameters, and the resulting moduli are presented in inset in Fig. 3.13. The graph demonstrates a dependence of the modulus on the fiber diameter similar to the one obtained by experiments of tensile tests of individual nanofibers [12]. The obtained data allow one to estimate the maximal elastic deformations of tested fibers both at equilibrium and during oscillations. It turned out that these deformations do not exceed 0.5 μm, which corresponds to a strain of about 0.2%. Therefore, apparently no plastic deformation occurs during the test. Also, note that high repeatability was achieved for a series of experiments on the same specimen lasting long periods of time (up to 1.5 months).

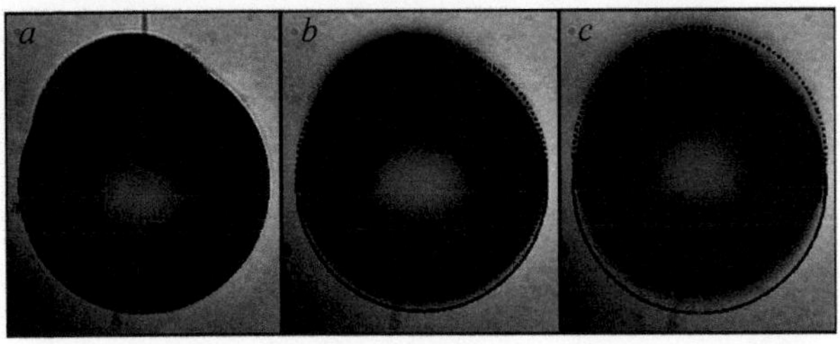

Figure 3.12 Series of still images (×1000 magnification, 10 sec exposure) of a ball's flights obtained for different input signal frequencies ($\nu = \Omega/2\pi$) with amplitude $A \approx 150$ nm. (a) $\nu = 0$ kHz; (b) $\nu = 1.1$ kHz; (c) $\nu = 1.8$ kHz (resonance frequency). The solid line corresponds to the ball location without flights, while the dotted line shows the average ball flight height. Reprinted from Ref. [34], with the permission of AIP Publishing.

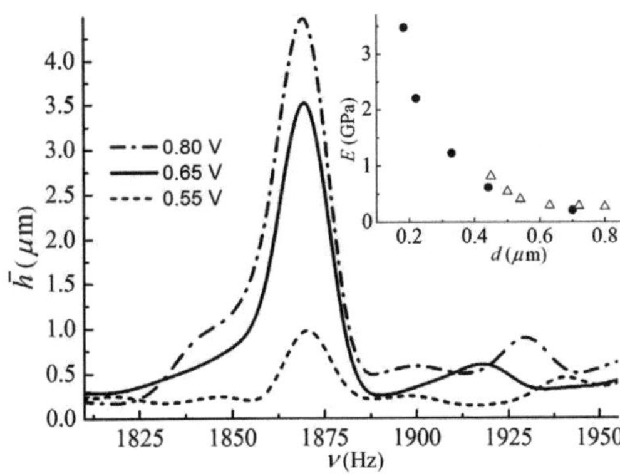

Figure 3.13 Typical plots of ball flight heights, $\bar{h}$, versus input signal frequency, $\nu = \Omega/2\pi$, for different input signal amplitudes. The inset demonstrates Young's modulus versus nanofiber diameter obtained with the help of the method described (dots) and with the help of the tensile method (triangles) [12]. Reprinted from Ref. [34], with the permission of AIP Publishing.

In summary, the described method for characterization of the mechanical properties of individual polymer nanofibers demonstrated its efficiency. It is based on the resonance of longitudinal forced oscillations of a bob suspended by a nanofiber that provides high accuracy of determined values of Young's modulus. The method is non-destructive and allows testing with various actuation parameters on the same specimen. Also, measurement of the mechanical properties of the sample while controlling the ambient environment (e.g., temperature, humidity, etc.) can be easily implemented.

Part II

Nanofiber Fabrication

Chapter 4

Electrospinning of Polymer Nanofibers

Polymer nanofibers can be fabricated from a semi-dilute polymer solution [52] with the help of the electrospinning process during which a polymer solution is extruded from a spinneret and by a strong electric field, a jet is formed at the tip (see famous Fig. 4.1, which has been published in many papers devoted to electrospinning). The resulting jet then undergoes extreme elongation and thinning, in the order of 10^5, accompanied by a strain rate of the order 10^3 s^{-1}. Extremely rapid solvent evaporation occurs with continued jet flow, resulting in the polymer solidification within a few milliseconds. Note that such significant stretching (up to 10^5 times) leads to stretching and orientation of the polymer chains. As a result, a non-equilibrium state of nanofiber polymer matrix is being formed, and this nonequilibrium state can be frozen due to very high rate of solvent evaporation. At the same time, possible relaxation processes are to be taken into account.

The physical principles of electrospinning are based on the Taylor examination, related to a specific problem of fluid mechanics [35–37]. In particular, Taylor showed that a viscous fluid is in equilibrium under an electric field when a jet leaving an orifice has the form of a cone (so-called Taylor cone) with a semi-vertical angle equal to 49.3° [38]. The initial phase of the thin fiber formation is shown in Fig. 4.2. The fluid goes out of a nozzle, forming the Taylor cone, which, in turn, is transformed via the transition zone into a jet.

Electrospun Polymer Nanofibers
Arkadii Arinstein

ISBN 978-981-4745-27-7 (Hardcover), 978-1-315-36462-9 (eBook)
www.panstanford.com

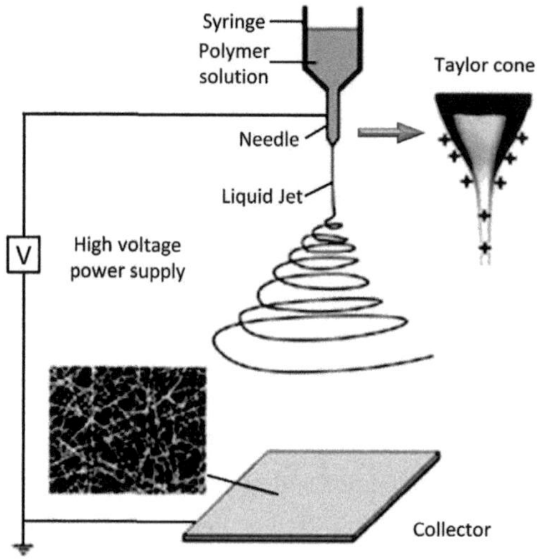

Figure 4.1 Schematic picture of the electrospinning process. The inset shows a schematic image of the Taylor cone.

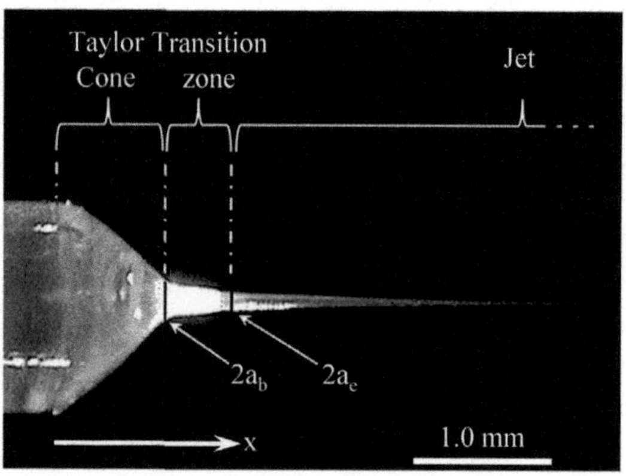

Figure 4.2 Taylor cone zone, with the transition zone, and the beginning of the thin jet in an electrically driven jet. The cross-sectional radii at the beginning and end of the transition zone are denoted by a_b and a_e, respectively. Reprinted from Ref. [58], Copyright 2008, with permission from Elsevier.

Nowadays, the necessary properties of spinnable polymer solution, as well as the process conditions that provide its stability, are well established. For this reason, the basics of electrospinning process of polymer nanofibers will be discussed very briefly. In the case of need, a detailed information can be found in many papers or reviews (see, for example, [39–46])

4.1 Spinability of Polymer Solution

It is well known that viscoelasticity is a prerequisite for polymer solution spinnability, meaning that spinnable solutions are semi-dilute, highly entangled. It turns out that electrospinning can result in polymer nanofibers by the use of only semi-dilute entangled polymer solution when $c/c^* > 3$. The use of lower polymer concentration results in the formation of aerosol particles. For semi-dilute entangled polymer solution, the following dependence of the fiber diameter on polymer concentration was observed (see Fig. 4.3) [52].

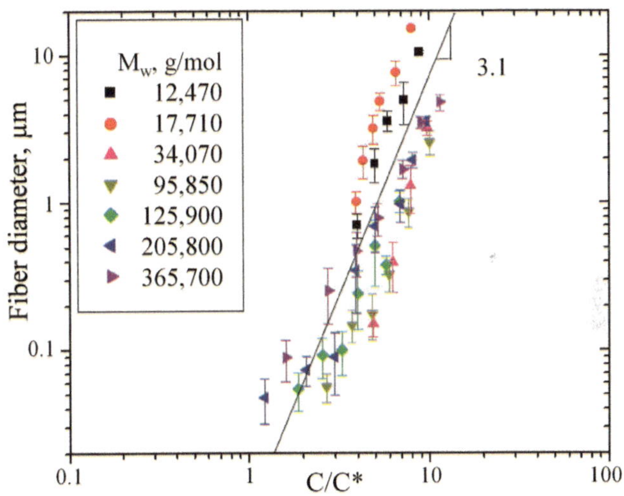

Figure 4.3 Variation of fiber diameter versus relative concentration, c/c^*, for different molecular weights M_w of polymethyl methacrylate (PMMA). Reprinted from Ref. [52], Copyright 2005, with permission from Elsevier.

Based on these experiments, the following scaling relationships for the dependencies of the fiber diameter D on the relative concentration c/c^* can be derived

$$D \propto (c/c^*)^{3.1} \tag{4.1}$$

which agrees with other measurements using a different type of polymer, resulting in $D \propto c^3$ [56] or $D \propto c^{2.99}$ [23].

The dependence of the zero shear rate viscosity η_0 of the polymer solution on the relative solution concentration c/c^* (see Fig. 4.4) exhibits three distinct concentration regions, identified as dilute ($c/c^* < 1$), semi-dilute unentangled ($1 < c/c^* < 3$), and semi-dilute entangled ($c/c^* > 3$) (see Fig. 4.4). Combining the dependences, shown in Figs. 4.3 and 4.4, one can get the following dependence of fiber diameter on polymer solution viscosity η_0:

$$D \propto \eta_0^{0.72} \tag{4.2}$$

A semi-dilute entangled polymer solution ($c/c^* > 3$) contains entangled macromolecules, ensuring the elastic behavior of the jet and continuous nanofibers. Conversely, in a semi-dilute unentangled solution, the polymer chains are not sufficiently entangled, resulting in beads rather than continuous nanofibers.

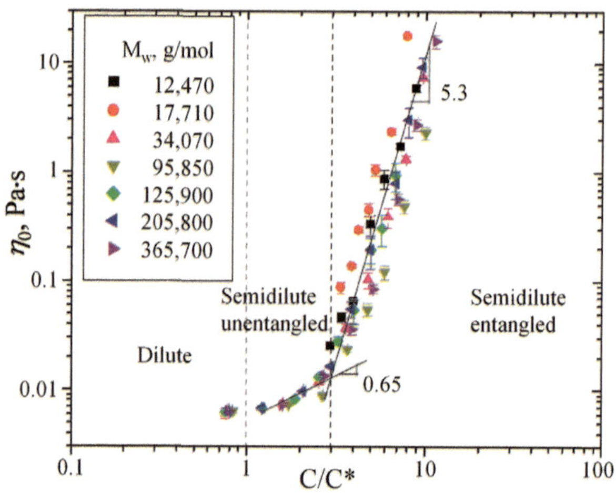

Figure 4.4 Zero shear rate viscosity η_0 versus relative concentration, c/c^*, for different molecular weights M_w of polymethyl methacrylate (PMMA). Reprinted from Ref. [52], Copyright 2005, with permission from Elsevier.

Now it is quite appropriate to ask: What will happen if, for the spinning, we use a polymer solution with relatively low concentration (semi-dilute unentangled solution with $c/c^* < 3$, and even dilute solution with $c/c^* < 1$)? The answer is: It turned out that electrospinning of dilute solution ($c/c^* < 1$) results only in the formation of aerosol (small separate particles), whereas electrospinning of semi-dilute unentangled solutions ($1 < c/c^* < 3$) results in the formation of stretched beads (see Fig. 4.5a). Further increase in the polymer concentration up to some critical value results in the initiation of the fiber formation (see Fig. 4.5b). The good defectless fibers can be obtained as a result of electrospinning only of highly entangled solutions with concentrations exceeding the next critical value (see Fig. 4.5d). And for the interstitial solution concentration, beaded fibers together with defectless fibers arise (see Fig. 4.5c). In other words, during the electrospinning of polymer solutions, an increase in polymer concentration results in the following modification in observed fiber morphology: beads only; beads with incipient fibers; beaded fibers; fibers only; and (in some cases) globular fibers/macrobeads [53]. Such a progression was observed by Shenoy et al. in [53] for four polymer systems: polystyrene, poly(ethylene oxide), poly(L-lactic acid), and poly(vinylpyrrolidone) (PVP). In Fig. 4.5, the aforementioned morphologies corresponding to different concentrations of spinning solution are presented for electrospun PVP.

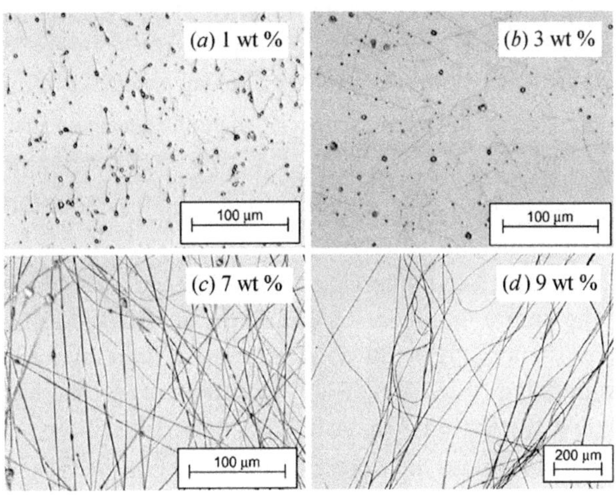

Figure 4.5 Optical microscopy of structures obtained by electrospinning PVP/EtOH solutions at different concentrations of PVP (M_w = 1300 k). 1 wt %: elongated beads (a); 3 wt %: incipient fiber formation (b); 7 wt %: beaded fibers and fibers (c); and 9 wt %: fibers only (d). Reprinted from Ref. [53], Copyright 2005, with permission from Elsevier.

For the quantitative characterization of the observed regularity, the number of entanglements per macromolecule, $(n_e)_{soln}$, can be introduced as the system parameter. The values of this parameter can be calculated with the help of the following empirical equation, which results in good estimation over a wide range of solution concentrations:

$$(n_e)_{soln} = \frac{M_w}{(M_e)_{soln}} = \frac{\phi_p M_w}{M_e}, \qquad (4.3)$$

where ϕ_p is the polymer volume fraction in polymer solution; M_w is the weight-average molecular weight of polymer; and M_e and $(M_e)_{soln}$ are the molecular weights of the average mesh between two adjacent entanglement junctions (topological knots) in the polymer melt and in the polymer solution, respectively. Thus, the above critical concentration values, ϕ_p^{cr}, correspond to the critical values of the entanglement parameter, $(n_e^{cr})_{soln}$.

It turned out that according to the experimental data obtained for all analyzed systems, these critical values of the entanglement parameter, $(n_e^{cr})_{soln}$, practically coincide for different polymers. And this fact, being of great importance, demonstrates that just the structure of the topological network in the spinning polymer solution affects the final morphology of electrospun nanofibers.

The first critical value, $(n_e^1)_{soln}$, which corresponds to the initiation of the fiber formation, is of about 2. And the second critical value, $(n_e^1)_{soln}$, which corresponds to the defectless fiber formation, amounts to about 3.5.

The typical picture demonstrating the effect of the entanglement level of the spinning polymer solution on the final morphology of electrospun nanofibers for the poly(lactic acid)/dimethylformamide/dichloromethane (PLA/DMF/DCM) system is presented in Fig. 4.6. On this plot, the dotted and dashed lines indicate the first and the second critical values of the entanglement parameter, $(n_e^{1,2})_{soln}$, respectively; the region between these lines corresponds the mixture of fibers and beads; and two sloped solid lines represent the samples with 670 and 190 k [53].

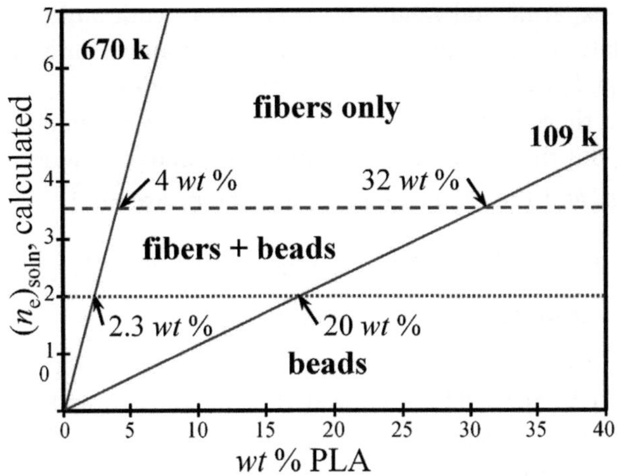

Figure 4.6 Plot of the calculated entanglement number as a function of concentration for PLA/(DMF/DCM) systems. The dashed line indicates the transition for complete fiber formation, i.e., $\left(n_e^2\right) = 3.5$, while the dotted line indicates the boundary between beads and a mixture of fibers and beads (shown in fill), i.e., $\left(n_e^1\right) = 2$. The solid lines represent the samples with M_w = 670 and 190 k, and arrows indicate the onset of fiber formation 2.3 and 18.5 wt % and complete fiber formation (4 and 31 wt %) for 670 and 190 k samples, respectively. Reprinted from Ref. [53], Copyright 2005, with permission from Elsevier.

In conclusion, note that in our opinion, the calculations of the entanglement parameter based on the Eq. (4.3) seem underestimating. Indeed, according to data of Shenoy et al. [53], the first entanglement parameter should amount to about 4. But, based on the equation for an average mesh size, ξ_0, of the network in a semi-dilute polymer solution (see Section 5.1 "Initial state of entangled semi-dilute polymer solution," Eq. (5.1)), the number of entanglements per macromolecule should be estimated as

$$\left(n_e\right)_{soln} = N / N_s \propto N(c / c^*)^{\nu/(1-3\nu)}, \tag{4.4}$$

where the crossover concentration of macromolecules overlap, c^*, depends on the macromolecule molecular weight, M_w, as

$$c^* \propto M_w^{3\nu-1} \tag{4.5}$$

Note that the dependence (19) is nonlinear, as distinct from Eq. (4.3), used by Shenoy et al. [53].

The experimental values of the crossover concentration, c^*, for PLA solution in a good solvent (benzene), obtained, for example, by Ren et al. [54], allows one to determine the index in Eqs. (19) and (20) and, thereafter, to estimate the entanglement parameter, $(n_e)_{soln}$. It turns out that the index $\nu \approx 0.59$ corresponds to values for good solvents; and the entanglement parameters are $\left(n_e^1\right)_{soln} \approx 11$ and $\left(n_e^1\right)_{soln} \approx 21$ for the molecular weight 190 k, and $\left(n_e^1\right)_{soln} \approx 4$ and $\left(n_e^1\right)_{soln} \approx 9$ for the molecular weight 670 k.

Note that, according to the measurements of Vega-Gonzalez et al. [55], DCM, used by Shenoy et al. [53], is also a good solvent for PLA, and the crossover concentrations, c^*, of PLA solution in both solvents are similar for similar molecular weights. Therefore, the estimations based on the data of Ren et al. [54] are suitable for the system examined by Shenoy et al. [53].

Perhaps, the aforementioned calculations result in an overestimation and the real values are lower, but, in any case, the entanglement level in the polymer solution, electrospinning of which results in the formation of a mixture of beaded fibers and beads, is high enough. Therefore, the connectivity breaking of the topological macromolecule network in such a system is impossible without intensive disentanglement process (see Section 5.3 "Disentanglement of polymer topological network in the further stages of electrospinning").

4.2 Hydrodynamics of Electrospinning

Polymer-based nanofibers are being fabricated with the help of electrospinning during which a polymer solution is extruded from a spinneret, and in the presence of a sufficiently strong electric field, a jet is formed at the tip. This jet then undergoes extreme elongation, thereby stretching the polymer molecules within it. As this occurs, the rapid solvent evaporation (about 1–10 msec) fixes the polymer matrix in this stretched, yet nonequilibrium state.

Under the electric field, the solution jet moves with acceleration so that the local velocity contains both longitudinal and radial components [57]:

$$v_z \approx v_0 \left(1+\frac{z}{z_0}\right)^2 + ..., \quad v_r \approx -v_0 \left(1+\frac{z}{z_0}\right)\frac{r}{z_0} + ..., \tag{4.6}$$

where v_0 is the jet initial velocity, and r_{J0} is the jet initial radius. The characteristic length z_0 determines the scale of velocity increase and depends on the flow rate Q, viscosity η, surface tension γ, and electric conductivity K of the solution, as well as on the applied electric field.

In a stable electrospinning process, flow rate, Q, should be sufficiently low, so the fluid motion is inertia less because the Reynolds number is sufficiently small ($\mathrm{Re} \sim \rho\eta Q/r_0 << 1$). Further, the diameter of a spinning jet is very small; therefore, we have to conclude that the surface tension should play a significant role in shaping the surface of the jet. Indeed, in a stable electrospinning process, the capillary number $Ca \sim \eta Q / r_0^2 \gamma$ is also sufficiently small (typically in the electrospinning $Ca \sim 10^{-2}$), and electrospinning occurs in a capillary-dominated regime. For example, electrified jets generated with $0.0063 < Ca < 0.018$ and large permeability, ε, are hyperbolic shaped with radius

$$r_J(z) \approx \frac{r_0}{1+z/z_0}, \tag{4.7}$$

The jet shape in the region near the orifice is elongated and slightly dependent on Ca, whereas the straight section of the jet demonstrates tighter dependence, where thickness increases with Ca for a fixed z (see Fig. 4.7).

This theoretical result is in good agreement with experimental observations. Several optical methods have been used to characterize electrospinning jets, ranging from simple low speed or static white light imaging of the jet profile near the tip to more complex techniques such as high-speed bright-field imaging of unique formations in the jet, and laser Doppler velocimetry, fluorescence microscopy, etc.

For example, the velocity measurement presented by Han et al. [58], is demonstrated in Fig. 4.8. The liquid solvent accelerates rapidly, moving away from the beginning of the jet, and at the distance 6 mm from the starting point, its velocity achieves 1 m/sec. In doing so, the stretching rate in the transition zone, dV/dx, lies in the range 100–1000 s^{-1}. The jet diameter also decreases very sharply, and it is too small for a measurement at distances exceeding 3 mm (see Fig. 4.8).

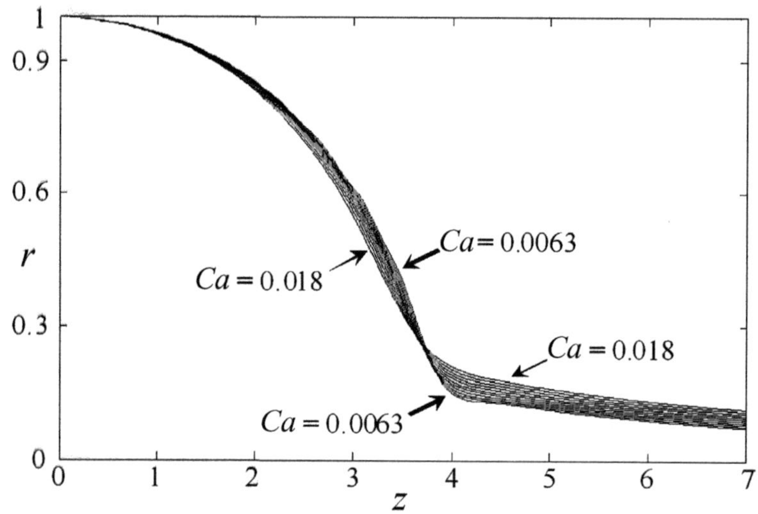

Figure 4.7 Jet shapes for various capillary numbers *Ca*. Reprinted from Ref. [57], Copyright 2010, with permission by the American Physical Society (https://doi.org/10.1103/PhysRevE.81.026313).

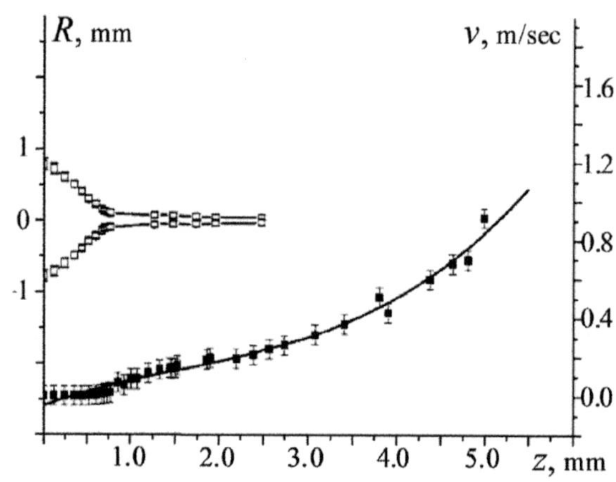

Figure 4.8 Profile of the transition zone following the modified Taylor cone and the beginning of the jet. The cross-sectional radius is shown as $R = R(z)$. Reprinted from Ref. [58], Copyright 2008, with permission from Elsevier.

Analogous results were obtained by Bellan et al. [59] (see Fig. 4.9): At the distance 1 mm from the starting point, the solvent

velocity achieves 0.25 m/sec, and such acceleration corresponds to the stretching rate of about 300 s^{-1}.

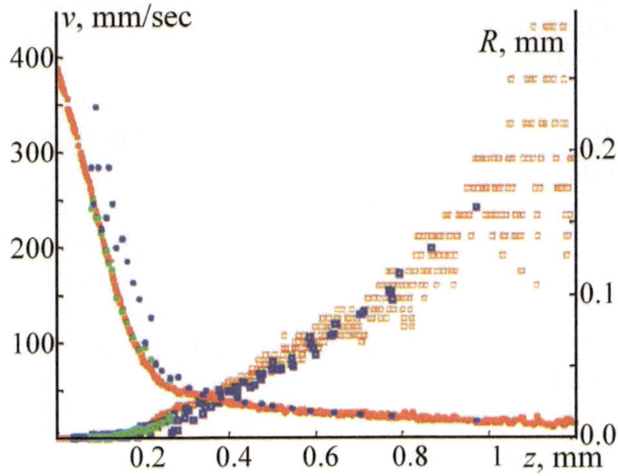

Figure 4.9 Velocity and jet radius plotted as a function of *z* position (length along the jet). Reprinted from Ref. [59], with the permission of AIP Publishing.

It is clear that the aforementioned velocity profile (4.6) demonstrating rapid unlimited increase in the longitudinal velocity is applicable only in the beginning of the spinning jet. The point is that along the jet, the solution properties are substantially being modified. Due to solvent evaporation, the solution viscosity sharply increases, and in the end, the jet solidifies. The stretching of the jet also affects the viscoelasticity of the polymer solution, increasing its viscosity. Note that the solution elasticity can also dramatically increase, resulting in the high-elastic state (see Section 5.2 "Stretching of polymer topological network in initial stage of electrospinning"). As a result, the increase in the longitudinal velocity v_z slows down, and finally it tends to saturation:

$$v_z \to v_\infty \tag{4.8}$$

In this case, the radial velocity v_r tends to zero, whereas the jet radius stops to decrease and tends to its final value $r_J^{(\infty)}$. Nevertheless, in the initial stage, Eq. (4.6) can be assumed as a satisfactory approximation, which allows one to estimate the evolution of the polymer solution state during electrospinning.

4.3 Hydrodynamic Instabilities of Spinning Jet

As mentioned earlier, during electrospinning, the jet flow is realized in a viscous (inertia less) capillary-dominated regime. The competition of surface forces and viscosity is a driving force for the development of an instability that can result in fiber failure caused by a jet fracture due to capillary wave breakup[1]. A fiber failure caused by breakage of the solidified jet due to the stresses overcoming some limiting tensile strength (necking, rupture) can also occur (this type of fiber failure will be discussed in Section 5.3.1 "Multiple necking").

Now we would like to focus on the aforementioned familiar capillary instability. In extreme cases, this capillary instability causes a cylindrical jet of liquid to collapse into separated droplets, but beaded fibers can also be formed.

The physical reason resulting in capillary instability can be understood from Fig. 4.10.

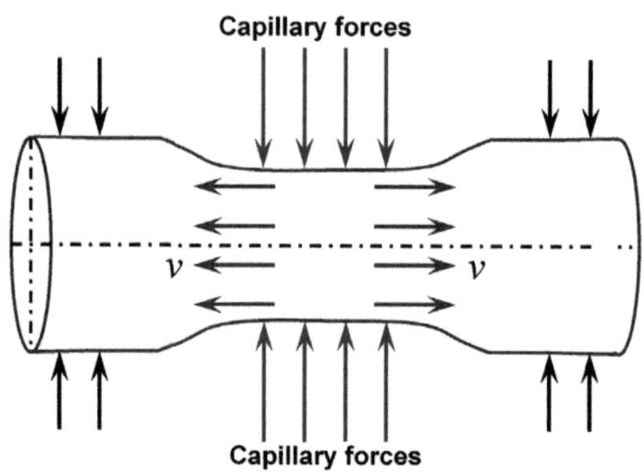

Figure 4.10 The physical mechanism resulting in unlimited growth of a fluctuation in the jet diameter.

If, for example, the jet diameter decreases due to a fluctuation, the capillary force extrudes fluid from the thinning region, decreasing its diameter; as a result, the fluctuation grows. Indeed, the capillary pressure P_C is inversely proportional to the jet radius

$$P_C = \frac{\gamma}{r}, \tag{4.9}$$

[1]Such type of instability is called capillary or Rayleigh instability.

and increases with decrease in the jet diameter (here γ is the surface tension, r is the radius of a jet). As a result, the capillary pressure in the fluctuating region is higher compared to the undeformed regions. The pressure difference is

$$\Delta P_C = \frac{\gamma \delta r}{r^2}, \tag{4.10}$$

and this difference results in the extrusion of a fluid and in the unlimited growth of a fluctuation.

In the opposite case, if a fluctuation results in an increase in the jet diameter, similar speculations show that the diameter of this fluctuation will grow unlimitedly. In doing so, there are no forces stopping an initial growth of small fluctuations in the form of the jet of a viscous liquid demonstrating no (or very low) elasticity.

The situation can dramatically change in the case of a semi-dilute entangled polymer solution where a polymeric network is formed. The variation in osmotic pressure can compensate for decrease in the capillary force, and the fluctuation dissipates (see Fig. 4.10).

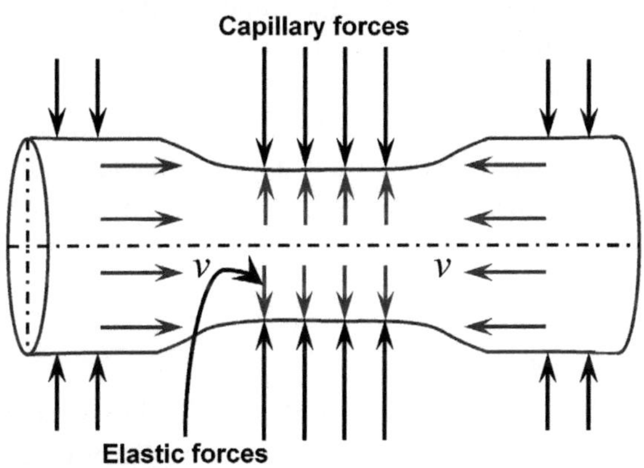

Figure 4.11 The physical mechanism resulting in the stabilization of the jet due to osmotic pressure of the polymer network.

Indeed, the pressure difference arising as a result of fluctuation in capillary radius is proportional to a radius variation (see Eq. 4.10), whereas the variation in the osmotic pressure having the opposite sign compared to capillary pressure is also proportional to a radius variation:

$$\Delta\Pi = -\frac{\partial\Pi}{\partial\xi_0}\delta\xi_0 = \frac{3k_\mathrm{B}T}{\xi_0^3}\frac{\delta r}{r}, \tag{4.11}$$

where ξ_0 is the mesh size of a polymeric network, $\delta\xi_0 = \xi_0\delta r/r$, T is the temperature, and k_B is the Boltzmann constant.

The fluctuation dissipates and the jet remains stable if the variation in the osmotic pressure $\Delta\Pi$ is dominant compared to the difference in the capillary pressure ΔP_C, i.e., the following inequality is satisfied

$$\frac{3k_\mathrm{B}T}{\xi_0^3} > \frac{\gamma}{r} \tag{4.12}$$

Note that in the case of the opposite inequality ($3k_\mathrm{B}T/\xi_0^3 < \gamma/r$), the capillary instability occurs.

Besides suppression of the capillary instability, the polymer network in the semi-dilute entangled polymer solution under high stretching can form filament-like structures providing the wholeness of a spinning jet, even in the case of development of capillary instability (see Section 5.2 "Stretching of polymer topological network in initial stage of electrospinning").

The temporal evolution of a viscoelastic liquid jet depends on the relation between the viscous, inertial, and elastic stresses and the capillary pressure [47]. In order to study this inertio-elasto-capillary balance in detail, two dimensionless parameters are defined. First of them is the Deborah number

$$De = \tau_\mathrm{El}\sqrt{\frac{\gamma}{\rho r_0^3}}, \tag{4.13}$$

defined as the ratio of the timescale for elastic stress relaxation, τ_El, to the "Rayleigh timescale" for inertio-capillary breakup of an inviscid jet, $\tau_\mathrm{R} = \sqrt{\rho r_0^3 / \gamma}$.

The second one is the Ohnesorge number

$$Oh = \frac{\eta_0}{\sqrt{\rho\gamma r_0}}, \tag{4.14}$$

which is the inverse of the Reynolds number based on a characteristic capillary velocity γ/η_0. In these expressions, ρ is the liquid density, η_0 is the liquid zero shear viscosity, γ is the surface tension, r_0 is the initial radius of the jet, and t_{El} is the relaxation time associated with the polymer solution.

The jet stability or the kinetics of instability development should be analyzed only in the framework of nonlinear models [48]. A low elasticity of a jet fluid (for example, a dilute polymer solution) can only affect the kinetics of instability development, but cannot result in jet stabilization [49, 50].

Nevertheless, the initial growth of the disturbances can be predicted using linear instability analysis for small perturbations. The linearization of the nonlinear equations describing the phenomenon in question results in the estimation of the characteristic time of the instability development, τ_{inst}, as well as of a scale of the appearing beads. All details can be found in many textbooks (see, for example [51], Chapter 1. "Capillary instability of free liquid jets"). We would like to restrict ourselves only to the final outcome.

Assuming a periodic disturbance to the radius along the z-axis (the direction of the jet flow), the perturbations of the jet radius, $\delta r(z, t)$, can be written in the following form (in the linear approximation, the amplitude of the radius perturbation should be much smaller, then the jet radius disappears):

$$r(z,t) - r_0 \equiv \delta r(z,t) \propto \exp(ikz + \omega t), \tag{4.15}$$

where $i = \sqrt{-1}$, $k = 2\pi/\lambda$ is a real positive wave number, with λ being the wavelength; and $\omega = \omega_r + i\omega_i$, with $\omega_r = 1/\tau_{\mathrm{inst}}$ being the growth rate of the disturbance, and ω_i being the oscillation frequency. Note that the characteristic time for the instability development $\tau_{\mathrm{inst}} = 1/\omega_r$ is associated to Rayleigh timescale, τ_{R}, defined earlier.

Simple calculations result in a characteristic relationship between the growth rate ω and the wave number k determining the perturbation spectrum:

$$\omega^2 = \frac{\gamma k}{\rho r_0^2}\left(1 - k^2 r_0^2\right)\frac{I_1(kr_0)}{I_0(kr_0)} \tag{4.16}$$

For a better understanding of the physical situation, in place of this well-known relationship (4.16), we will consider the characteristic time of the instability development, τ_{inst}, depending on the scale of an unstable perturbation, which cannot be very small: $\lambda > 2\pi r_0$ (see Fig. 4.12):

$$\tau_{\text{inst}} = \tau_0 \sqrt{\frac{I_0(x)}{x(1-x^2)I_1(x)}}, \tag{4.17}$$

where $x = kr_0 = 2\pi r_0/\lambda$, $\tau_0 = \sqrt{\rho r_0^3 / \gamma}$. The traditional dependence of the growth rate ω on the wavenumber k for the Rayleigh–Plateau instability (4.16) is presented in this figure as inset.

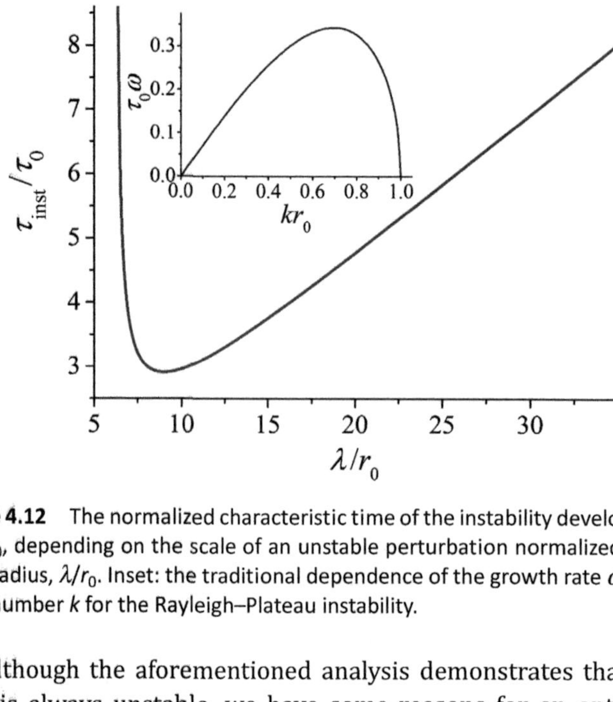

Figure 4.12 The normalized characteristic time of the instability development, $\tau_{\text{inst}}/\tau_0$, depending on the scale of an unstable perturbation normalized by the fiber radius, λ/r_0. Inset: the traditional dependence of the growth rate ω on the wavenumber k for the Rayleigh–Plateau instability.

Although the aforementioned analysis demonstrates that a jet flow is always unstable, we have some reasons for an optimism. Indeed, according to the aforementioned speculations, a perturbation of small amplitude starts to grow. But if the scale of this perturbation is too small (the minimum scale of an unstable perturbation should be larger than $2\pi r_0$) or too large, the waiting time of the instability development can prove to be longer than the process duration. The minimum waiting time $\tau_{\text{inst}}^{\min} \approx 2.9\tau_0$ corresponds to an "optimal" perturbation with the scale $\lambda \approx 9r_0$. Assuming $\rho \approx 10^3$ kg/m^3, $r_0 \approx 10^{-3}$ m, and $\gamma \approx (50 \div 100) \times 10^{-3}$ N/m, we get that $\tau_0 \approx (3.1 \div 4.5) \times 10^{-3}$ sec, and $\tau_{\text{inst}}^{\min} \approx (9.2 \div 13) \times 10^{-3}$ sec. This estimation shows that the minimum waiting time $\tau_{\text{inst}}^{\min}$ is comparable (a little smaller) than the duration of the electrospinning process, and the capillary instability can break up the spinning jet, resulting in aerosol formation. And, indeed, such a situation can often be observed in the real process. Nevertheless, there are mechanisms preserving the wholeness of an unstable jet. The point is that, as the local radius of constrictions in the jet decreases under the action of surface tension, elastic stresses grow and become comparable to the capillary pressure, leading to the formation of a uniform thread connecting two primary drops. Such a structure is called 'beads-on-a-string'; and a possibility of the appearance of this structure during the spinning process will be discussed in Section 5.2 "Stretching of polymer topological network in initial stage of electrospinning." According to Bhat et al., "the beads-on-a-string phenomenon occurs as a result of the delicate interplay of four forces: capillary, viscous, elastic and inertial, and when any of these forces dominates the others, it can suppress the bead formation" [47]. It turned out that the bead formation occurs at enough low (but non-zero) values of Deborah (4.13) and Ohnesorge (4.14) numbers ($De < 1$ and $Oh < 1$). In this case, the rapid relaxation of elastic stress does not prevent the perturbation growth (the bead growth), but, at the same time, viscous forces are comparable to the inertia and surface forces and such a balance prevents the full breakup of the jet. As a result, the beads-on-a-string structure is forming.

In the opposite case, when the relaxation of an elastic stress is slower than the rate of a deformation increase, elastic forces become dominant, and the bead formation can be suppressed.

Finally, it is suitable to mention other instability types (the detailed analysis of possible instabilities in spinning jet is presented by Reneker and Yarin in their review [46]). The first of them is the bending instability. After the jet flowed away from the droplet in a nearly straight line, it bent into a three-dimensional coil. After several turns were formed, a new electrical bending instability formed a smaller coil on a turn of the larger coil (see Fig. 4.13).

This type of instability arises due to the rapid growth of bending perturbations under the influence of the charge carried with the jet [42, 46]. The physical mechanism resulting in the growth of small bending perturbations that, in turn, result in the development of bending instability can be easily understood with the help of the inset (of the magnifying glass) in Fig. 4.13: in the case of inflection of a straight jet (even of a very small one), the Coulomb repulsion forces cannot compensate each other (they turned not parallel); and the resulting force is acting in the deflection direction, increasing the arising fluctuation.

Note that the aforementioned bending instability does not result in electrospun nanofiber failures and, from the point of view of the problem in question, has no interest for us. The other type of instability, namely, branching of spinning jet, which occurs under some circumstances, is also beyond the framework of our consideration.

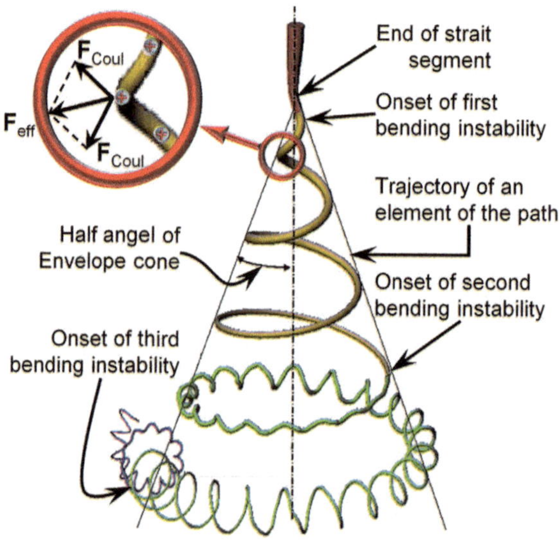

Figure 4.13 Schematic view of the bending instability in a spinning jet. The inset (circle representing a magnifying glass) explains the mechanism of the bending instability developing as a result of the action of uncompensated forces of Coulomb repulsion. Reprinted from Ref. [46], Copyright 2008, with permission from Elsevier.

4.4 Solvent Evaporation and Polymer Solidification

In spite of many speculations and estimations, the impact of the evaporation process on the physical features of the electrospun polymer nanofibers has not been studied in detail. The point is

that the rapid solvent evaporation in the final stage of the spinning process fixes the polymer matrix in nonequilibrium (stretched) state.

The theoretical examination of the solvent evaporation from a spinning jet is a very hard problem. Nevertheless, namely, theoretical analysis combined with computer simulations can clarify some aspects of the problem in question. For example, the role that the high evaporation rate plays in the fabrication of polymeric electrospun nanofibers has already been discussed by Koombhongse et al. [60] and treated quantitatively by Guenthner et al. [61]. In this paper, the system in question was analyzed numerically on the basis of the nonequilibrium thermodynamics with the help of nonlinear reaction–diffusion equation, more exactly, with the help of the time-dependent Ginzburg–Landau equation in combination with Flory–Huggins free energy of mixing and Fick's law for solvent evaporation.

For the required analysis, the system evolution equation, based on the the Ginzburg–Landau equation, was written as:

In this analysis, the time evolution equation for the volume fraction (volumetric concentration) of polymer, $\phi(\vec{r},t)$, was based on the Ginzburg–Landau equation:

$$\frac{\partial \phi}{\partial t} = \tilde{\nabla} \cdot \Lambda \tilde{\nabla} \frac{\delta G}{\delta \phi} \tag{4.18}$$

where $\phi(\vec{r},t)$ is the volume fraction (volumetric concentration) of polymer; the concentration-dependent mobility of the polymer solution, Λ, obeys Onsager reciprocity relation: $1/\Lambda = 1/\Lambda_1 + 1/\Lambda_2$ with $\Lambda_1 = \phi n D_1$ and $\Lambda_2 = (1 - \phi)\, nD_2$, $D_{1,2}$ are the self-diffusion coefficients of the polymer and the solvent, respectively, and n is the degree of polymerization of the polymer. The functional derivative is defined as $\delta/\delta\phi \equiv \partial/\partial\phi - \tilde{\nabla} \cdot \partial/\partial\left(\tilde{\nabla}\phi\right)$, and $\tilde{\nabla} \equiv \left(\vec{i}\partial/\partial x + \vec{j}\partial/\partial y + \vec{k}\partial/\partial z\right)$ is the standard spatial operator in the vector form.

The total free energy G contains both a local and a non-local free energy terms of a free energy density of the system:

$$G = \int \left(g_{\text{loc}} + g_{\text{non-loc}}\right) dV \tag{4.19}$$

In this equation (4.19), the local free energy density g_{loc} corresponds to the classical Flory–Huggins free energy of mixing

$$\frac{g_{\text{loc}}}{k_{\text{B}}T} = \frac{1}{n}\phi \ln\phi + (1-\phi)\ln(1-\phi) + \chi\phi(1-\phi), \tag{4.20}$$

and the non-local free energy density $g_{\text{non-loc}}$ is expressed in terms of volume fractions of the system components as:

$$\frac{g_{\text{non-loc}}}{k_{\text{B}}T} = \frac{1}{36}\left(\frac{a_{\text{p}}^2}{\phi} - \frac{a_{\text{s}}^2}{1-\phi}\right)\left|\tilde{\nabla}\phi\right|^2 \quad (4.21)$$

where k_{B} is the Boltzmann constant, T is the absolute temperature, χ is the Flory–Huggins interaction parameter, and $a_{\text{p,s}}$ are the scales of monomer (persistent lengths) of polymer macromolecules and solvent molecules, respectively.

And finally, zero-boundary conditions on the jet surface are assumed, since the solvent evaporates into infinite volume and its volumetric fraction in the surrounding air is negligibly small.

The above system of equations is very complicated and was analyzed numerically. In particular, it was demonstrated that in the case of relatively low solvent evaporation rate, i.e., of relatively slow solvent loss, a region enriched by polymer is being formed near the surface of the fiber. Moreover, with the shrinking of the spinning jet due to solvent loss, the polymer concentration at the inner side of this region decreases continuously toward the center of the fiber. Nevertheless, when the difference in the polymer concentration in the near surface region and in the fiber center is too high, the concentration gradient starts to decrease due to diffusive fluxes (see Fig. 4.14). As a result, the relative homogeneous fibers are formed.

The situation changes dramatically when the evaporation occurs very fast. In this case, a thin skin consisting of nearly pure polymer forms rapidly at the initial stage of the spinning. This skin layer of "solidified" polymer differs sharply from the fiber core throughout the evaporation process, demonstrating a high resistance to deformations (compared to solution). With time, the skin layer grows thicker, but the difference in polymer concentration in the solution on the inner skin surface and in the center of the fiber remains almost constant (see Fig. 4.15). In addition, the formed skin acts as a barrier, resisting further solvent evaporation. Thus, in the case of high evaporation rate, electrospinning results in the formation of sharply heterogeneous, porous fibers containing noticeable amount of residual solvent.

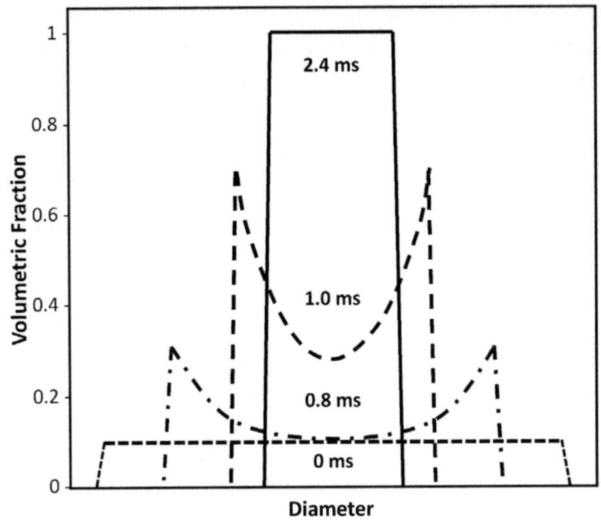

Figure 4.14 The evolution of the cross-sectional polymer concentration profile inside an emerging fiber in the case of relatively low solvent evaporation rate, i.e., of relatively slow solvent loss. Reprinted from Ref. [61], Copyright 2005, with permission from John Wiley and Sons.

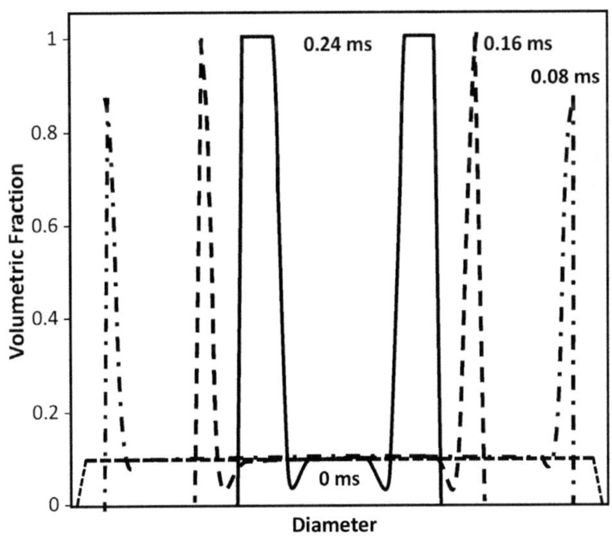

Figure 4.15 The evolution of the cross-sectional polymer concentration profile inside an emerging fiber in the case of high solvent evaporation rate, i.e., of rapid solvent loss. Reprinted from Ref. [61], Copyright 2005, with permission from John Wiley and Sons.

These results indicate that despite the rapid evaporation that has occurred, the collected electrospun nanofibers still contain a significant amount of solvent. The presence of the solvent, which now evaporates even slower due to the barrier that has formed, apparently results in " relaxation" of the fabricated nanofibers. This relaxation causes certain post-processes occurring within the system, and the effect of this residual solvent is crucial to understanding the mechanisms that affect the nanofibers' mechanical properties (see Section 4.5 "Post-fabrication relaxation").

A similar phenomenological model describing the solvent evaporation process was also proposed by Wu et al. [62] on the basis of Fick's law of mass diffusion and the following assumptions: (1) No temperature gradient exists inside the jet; (2) the mass diffusion and transfer are axisymmetric; (3) micro-convection due to concentration gradients is negligible; and (4) no chemical reactions occur inside the jet. The polymer–solvent interaction was taken into account according to the Flory–Huggins equation.

The governing equation of mass diffusion in radial direction of the jet is

$$\frac{\partial C_s}{\partial t} = \frac{1}{r}\frac{\partial}{\partial r}\left(Dr\frac{\partial C_s}{\partial r} \right), \qquad (4.22)$$

where D is the mass diffusivity and C_s is the solvent mass concentration. The mass concentrations of solvent (C_s) and polymer (C_p) and their specific partial volumes (V_s and V_p) obey the normalization condition $V_sC_s + V_pC_p = 1$.

Initial conditions at the beginning of the drying process are

$$C_s(t = 0, r) = C_{s0};\ R(t = 0) = R_0, \qquad (4.23)$$

where $R(t)$ is the current radius of the jet.

The boundary condition at the center of the jet is trivial: $\partial C_s / \partial r|_{r=0} = 0$, whereas the boundary conditions at the mobile surface of the jet are

$$\frac{D}{1 - V_sC_s}\left.\frac{\partial C_s}{\partial r}\right|_{r=R(t)} = -\frac{k_sM_s\left(P_s - P_\infty\right)}{R_gT}, \qquad (4.24)$$

$$\frac{\partial R(t)}{\partial t} = -\frac{k_sM_sV_s\left(P_s - P_\infty\right)}{R_gT}, \qquad (4.25)$$

where k_s is the coefficient of solvent mass transfer at the jet surface, M_s is the solvent molar weight, P_s is the solvent saturation vapor pressure near the jet surface, P_∞ is the solvent vapor pressure in the atmosphere far from the jet surface (it was assumed that $P_\infty = 0$), T is the absolute temperature near the surface, and $R_g = N_A k_B = 8.3144$ J/mol is the universal gas constant (N_A is the Avogadro constant and k_B is the Boltzmann constant).

The solvent saturation vapor pressure near the jet surface, P_s, can be estimated according to the Flory–Huggins equation:

$$P_s = P_{s0}\varphi_1 \exp\left(\varphi_2 + \chi\varphi_2^2\right) \tag{4.26}$$

where P_{s0} is the saturation vapor pressure of the pure solvent, $\varphi_{1,2}$ are the solvent and polymer volume fractions, respectively ($\varphi_1 + \varphi_2 = 1$), and χ is the Flory–Huggins polymer–solvent interaction parameter.

As in the previous case (analysis of Guenthner et al. [61]), the system of the aforementioned differential equations is also very complicated and was analyzed numerically.

The simulations have shown, in particular, that highly inhomogeneous concentration profiles are formed as a result of solvent evaporation in jets with radii of tens or a few micrometers, and that the degree of inhomogeneity decreases with decrease in the initial jet radius down to nano-sizes (see Fig. 4.16). Note that inhomogeneity in solvent distribution can affect the rheological behavior of a jet, and this fact should be taken into account in further studies.

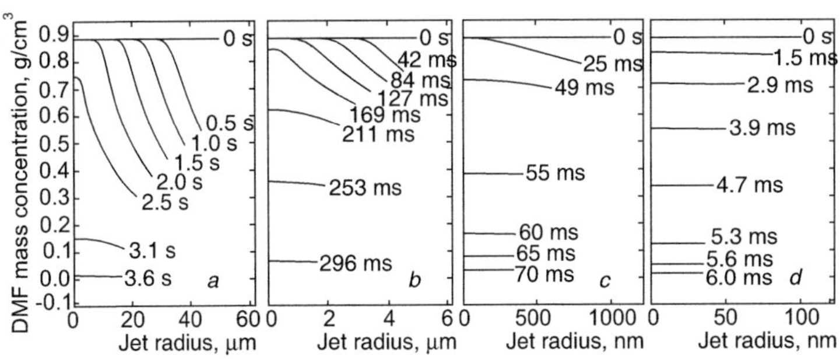

Figure 4.16 Evolution of the concentration profile of the solvent dimethylformamide (DMF) in jets with initial radii: $R_0 = 50$ μm (a), $R_0 = 5$ μm (b), $R_0 = 1000$ nm (c), and $R_0 = 100$ nm (d). Reprinted with permission from Ref. [62], Copyright 2011, AIP Publishing LLC.

The kinetics of the decrease in the jet radius occurring as a result of solvent evaporation is presented in Fig. 4.17.

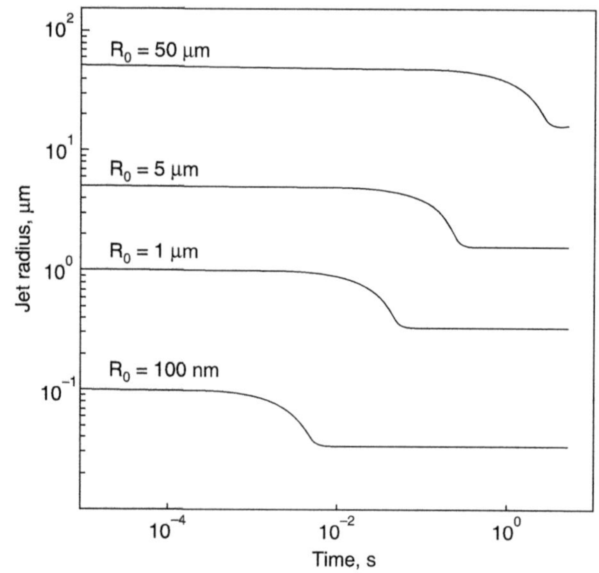

Figure 4.17 Time evolution of polymer jet radius as a result of solvent evaporation for initial jet radii: R_0 = 50 μm, R_0 = 5 μm, R_0 = 1000 nm, and R_0 = 100 nm. Reprinted with permission from Ref. [62], Copyright 2011, AIP Publishing LLC.

These data also allow to estimate the time required for the almost full solvent evaporation, and this evaporation time can be called "the drying time." It turned out that a rapid decrease in drying time from seconds for jets having radii of tens or a few micrometers, down to a few milliseconds for jets with submicron initial radius (see Fig. 4.18, as well as Figs. 4.16 and 4.17), which also allow to observe this regularity.

The calculated estimations of evaporation time seem quite reasonable, since the obtained values are in good agreement with the estimations of the spinning time based on the experimental data. Note that although the solvent evaporation rate is mainly determined by solvent properties, this rate depends also on the process conditions, in particular, on humidity of the surrounding atmosphere. Thus, Tripatanasuwan et al. show that the evaporation rate and solidification of the spinning jet are controlled by the relative humidity during electrospinning of poly(ethylene oxide), affecting the thickness of the obtained nanofibers [63]. As the relative humidity increased, the solidification process became slower, allowing elongation of the charged jet to continue longer and thereby to form thinner fibers.

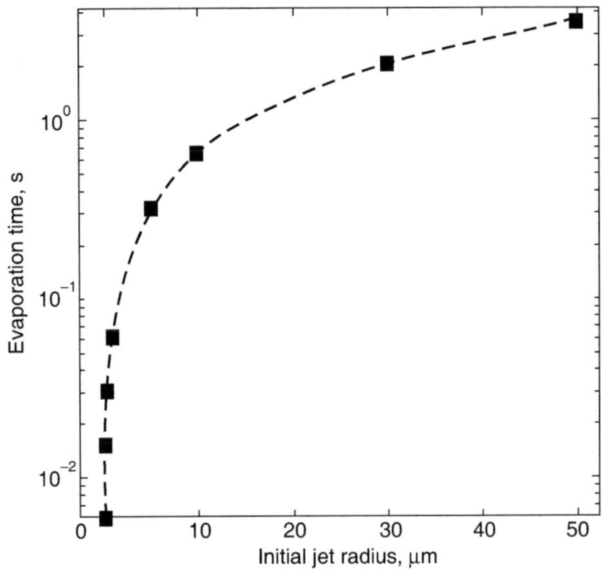

Figure 4.18 The drying time of the jets with different initial radii. Reprinted with permission from Ref. [62], Copyright 2011, AIP Publishing LLC.

Although the model of Wu et al. allows to correctly estimate the evaporation time, some system features of great importance are not taken into account. For example, the nonequilibrium state of the polymeric part of the system modifies the polymer–solvent interaction and due to this circumstance, the solvent saturation vapor pressure is also modified, affecting the solvent evaporation rate. Moreover, the modifications of the polymer state that occur during the electrospinning cannot be analyzed in the framework of this model. From this point of view, the concept of Dayal and Kyu, which takes into account the phase state of the system, is suitable to be mentioned [64]. Briefly, this concept assumes that the coupling of the phase separation dynamics with solvent evaporation through the fiber surface can result in the development of heterogeneous morphology (i.e., in the formation of supermolecular structures), such as polymer droplets, interconnected spinodal structure, modulations in the polymer density of electrospun nanofibers, including the formation of porous structures, etc.

Finally, note that in spite of all the discussed results, the process of solvent evaporation from the spinning jet requires additional, more detailed examination.

4.5 Post-fabrication Relaxation

As mentioned earlier, the investigation of the very rapid solvent evaporation occurring in the electrospinning process is a complex experimental problem. In place of this one, we focus on a special case of the system discussed by Guenthner et al. [61]. Their theoretical analysis and simulation have shown that the polymer molecules in the electrospun fibers are inhomogeneously distributed in the cross section of the nanofiber, forming a compact surface or "shell". Also, a spongy core is formed as a result of the entrapped solvent, which cannot quickly escape due to the low penetrability of the formed shell. Clearly, a tubular nanofiber filled with solvent can be regarded as a special case of the Guenther et al. system. As an experimental system, tubular nanofibers are suitable for use, thereby enabling observation of the decrease in the amount of solvent inside the nanofibers due to evaporation. In this manner, the kinetics of solvent evaporation through the shell can be tracked, which, in turn, allows one to estimate the duration of possible relaxation process. Such an examining was fulfilled by Arinstein and Zussman [65].

The method of fabricating such tubular nanofibers is called co-electrospinning. Briefly, two polymer solutions are supplied from a spinneret consisting of two coaxial capillaries. Close to the spinneret's exit, the emerging jet has a cross section composed of two concentric layers, which forms a core–shell structure (see Fig. 4.19).

The origin of the core–shell structure at the initial stage of co-electrospinning can be seen in Fig. 4.20.

The outer surface of the shell experiences a dry spinning process, as it is exposed to the surrounding air. Thus, the solvent diffusion and the evaporation at the fiber/vapor interface promote solidification and the resultant morphology at the outer surface. On the other hand, the inner layer of the shell experiences a wet spinning process, as it is in contact with the relatively nonvolatile core solution (e.g., aqueous solution), which is a non-solvent to the polymer shell. The core solution can then be regarded as a coagulation bath for the shell.

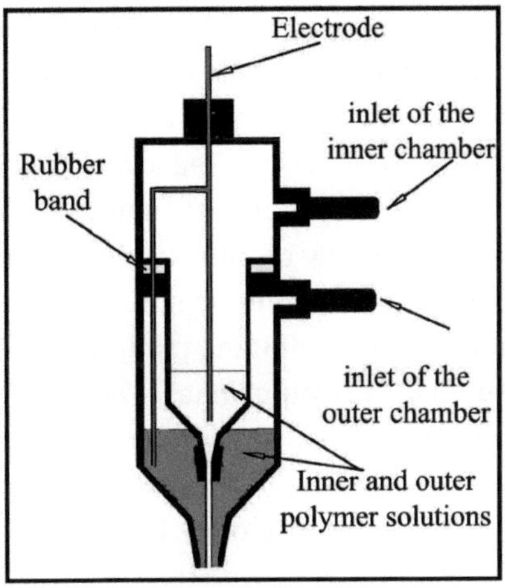

Figure 4.19 Schematic description of co-electrospinning equipment allowing one to fabricate nanofibers having a core–shell structure.

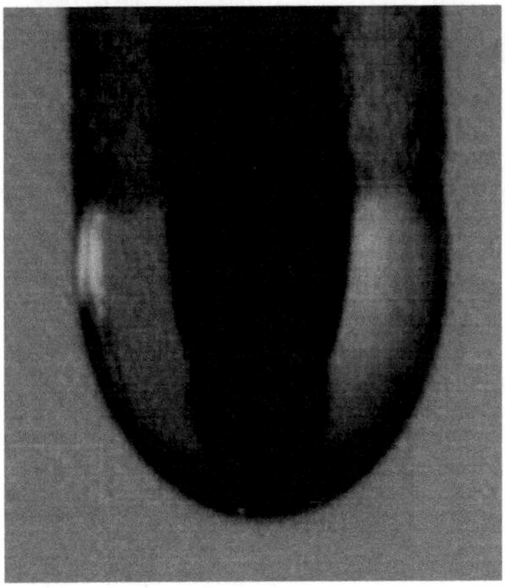

Figure 4.20 Origin of the core–shell structure at the initial stage of co-electrospinning.

4.5.1 Experimental Observations

First of all, let us consider an experiment in which manipulation with core–shell nanofibers allows one to collect the necessary experimental data describing the evaporation kinetics.

The collected tubular nanofibers with the entrapped core solvent were observed using an optical microscope immediately after their fabrication, enabling the experimental analysis of further evolution of the system in question. This process has a typical time duration of tens of seconds, which means that while the initial stage could not be observed, some deductions regarding it can be formulated from subsequent developments and an analysis of the entire process (see Section 4.6.2 "Buckling of tubular nanofiber"). Subsequent experimental observations of the as-spun nanofibers reveal that they consist of microtubes containing many slugs of entrapped core solvent that are bound by a vapor phase. A typical image of an as-spun fiber with an entrapped slug bound by two menisci is depicted in Fig. 4.21a. The observed slug shortens as evidenced by the menisci coming toward each other at a velocity V_m and finally disappears as a result of solvent evaporation (see Fig. 4.21b). Typical patterns of meniscus displacement and its velocity V_m are displayed in Fig. 4.22. Initially, the meniscus moves at a constant rate (see the inset in the graph for the pattern of a long slug). Then its velocity slows down, and finally, when the distance between the left and right menisci of the liquid is small, the meniscus displacement tapers off slowly. When the entire slug has eventually evaporated, we are left with a hollow.

It is clear that the menisci motion is caused by the evaporating solvent, so if a relationship between these processes can be found, it will assist in studying the evaporation kinetics of the as-spun nanofibers. Note that the typical evaporation time of the core solvent is in the order of tens of seconds. For this reason, along with the observations described earlier, one can conclude that relaxation and the further evolution of the polymeric matrix take place inside the nanofibers after their formation.

As the solvent evaporation results in a decrease in the liquid quantity inside the fibers, this has a direct correlation with the menisci motion. Therefore, by studying the menisci displacement pattern, one can clarify the processes accompanying the solvent evaporation.

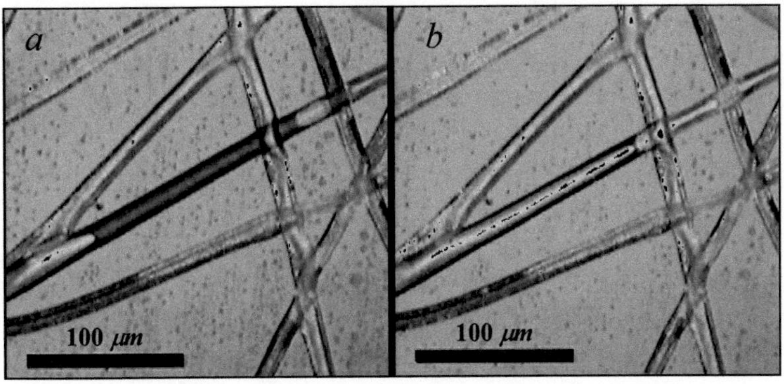

Figure 4.21 Typical image of as-spun core–shell fibers. One can see that one of them contains an entrapped slug bounded by two menisci (a); and disappearance of this slug as a result of solvent evaporation (b). Reprinted from Ref. [65], Copyright 2007, with permission by the American Physical Society (https://doi.org/10.1103/PhysRevE.76.056303).

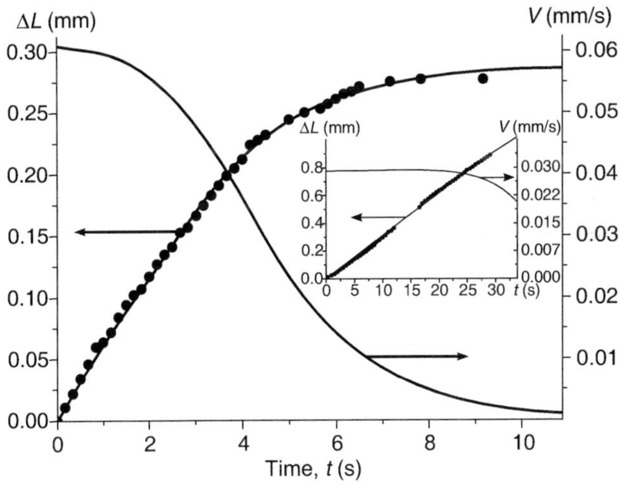

Figure 4.22 The meniscus displacement $\Delta L(t) = L_0 - L_{cap}(t)$ (left axis) and its rate (right axis) versus time. The dots display the experimental data, and the lines display the theoretical model [two-parameter fitting with the help of Eq. (4.40)]. The inset displays another set of experimental data for which evaporation through the capillary wall is dominant. Reprinted from Ref. [65], Copyright 2007, with permission by the American Physical Society (https://doi.org/10.1103/PhysRevE.76.056303).

4.5.2 Preliminary Estimates

A reasonable starting point is to consider the dominant processes affecting the evaporation kinetics (see Fig. 4.23 with a schematic description of this process). The evaporation of the core solvent can take place at both the shell and the menisci. At first glance, it would appear that the latter channel would be dominant, since the filtration of solvent through the capillary wall is limited by a very low diffusion coefficient ($D_s \propto 10^{-13} \div 10^{-12}$ m^2/sec [67]). However, taking into account the ratio of surface areas of the meniscus and capillary wall, it is found that solvent evaporation from the meniscus surfaces has a limited influence on the evaporation kinetics, and that solvent filtration through the capillary wall controls the process in question. Indeed, the characteristic time of the solvent evaporation from the meniscus, t_m, can be approximated as

$$t_m \propto D_g^{-1}\left(2L_0C_l / C_g\right)^2 \approx 4\times10^4 \text{ sec}, \tag{4.27}$$

where D_g ($D_g \propto 10^{-5}$ m^2/sec) and C_g are the diffusion coefficient and solvent density in the gas phase, C_l is the density of liquid solvent ($C_l/C_g \propto 10^3$), and L_0 is half of the initial length of the slug ($L_0 \approx 3.10^{-4}$ m).

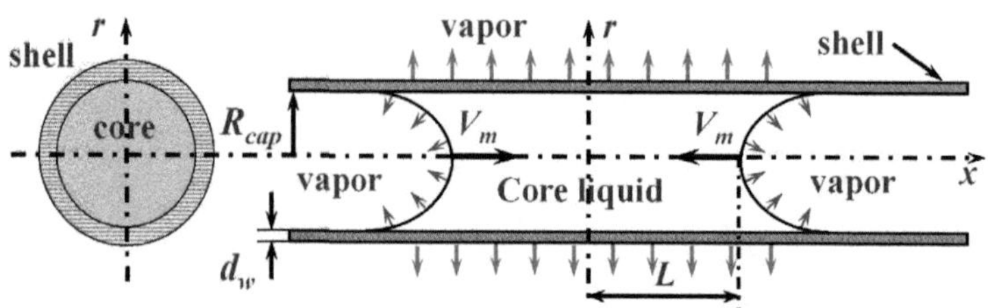

Figure 4.23 A scheme of the dominant processes determining the evaporation kinetic. Reprinted from Ref. [65], Copyright 2007, with permission by the American Physical Society (https://doi.org/10.1103/PhysRevE.76.056303).

This calculated time duration exceeds the experimentally observed time of the evaporation process by two and a half orders of magnitude. Even if such a difference seems to be negligible, one should take into account that the above value is underestimated. The reason is that the evaporated gas is assumed to be removed without any impediments. In reality, outgas can occur only through the tube shell, and if we take this into account, we come up with a dramatic increase in the evaporation time (and if the concentration of evaporated gas manages to equal the saturated concentration, the process can stop altogether).

However, the most cogent argument supporting the negligibility of solvent evaporation from the meniscus surfaces is that the evaporation time is dependent on the displacement rate of the menisci. According to experimental observation, if the length of a slug is long enough, the displacement rate of the meniscus is nearly constant (see the inset in Fig. 4.22) and decreases exponentially when the length of the slug becomes short enough (see Fig. 4.22). However, the solvent evaporation from the meniscus surfaces results in other dependence of type of inverse square root, $V_{\mathrm{m}}(t) \propto \sqrt{1/t}$. Therefore, the kinetics of the process in question cannot be explained by solvent evaporation from the meniscus surfaces, but only by infiltration of the solvent through the capillary wall and its subsequent evaporation at the fiber/vapor interface (the last process occurs very quickly).

The time necessary for the infiltration of the liquid solvent through the wall of a capillary can be determined by considering the flux of the liquid that read

$$2\pi R_{\mathrm{cap}} 2L_0 D_{\mathrm{s}} (fC_{\mathrm{l}}/fr) t_{\mathrm{w}} \propto 2\pi R_{\mathrm{cap}} 2L_0 D_{\mathrm{s}} (C_{\mathrm{l}} / d_{\mathrm{w}}) t_{\mathrm{w}} = \pi R_{\mathrm{cap}}^2 2L_0 C_{\mathrm{l}}, \tag{4.28}$$

where R_{cap} and d_{w} are the radius of a capillary and thickness of its walls, respectively.

Assuming $D_{\mathrm{s}} \propto 10^{-13}$ m^2/sec, $R_{\mathrm{cap}} \propto 10^{-5}$ m, and $d_{\mathrm{w}} \propto 0.5 \times 10^{-6}$ m,

$$t_{\mathrm{w}} \propto R_{\mathrm{cap}} d_{\mathrm{w}}/D_{\mathrm{s}} \approx 1 \text{ min} \tag{4.29}$$

This timescale is in good agreement with the experimentally observed duration of the evaporation process. Therefore, we can conclude that the dominant process determining the rate of menisci recession is infiltration through the wall of the capillary by the liquid solvent and its subsequent evaporation. Therefore, from this point on, we will deal exclusively with this process and disregard evaporation through the menisci.

4.5.3 Kinetic Equations Describing the System Evolution

Now our aim is determining the entrapped slug's evolution and its evaporation through the shell over time.

According to experimental observations, no deformations of most of capillaries accompany the solvent evaporation. Hence, the pressure inside the capillary decreases, and this gradient of pressure results in a solvent flow directed from the meniscus to the center of the slug, compensating the solvent evaporation inside the capillary. Since the length of the slug is much larger than its radius, $L>>R_{cap}$, and Reynolds number, $Re = C_l R_{cap} V_m/\eta \propto 10^{-6}$, is very low, locally we can assume Poiseuille's flow

$$V(x,t) = -\frac{R_{cap}^2}{8\eta}\frac{\partial P(x,t)}{\partial x}, \tag{4.30}$$

where $V(x, t)$ is the mean velocity of solvent in cross section of capillary and η is the solvent viscosity.

The second equation describing the process incorporates the mass conservation law, which takes into account the solvent evaporation,

$$\frac{\partial V(x,t)}{\partial x} = -\frac{2D_s}{R_{cap}d_w}\cdot\frac{1+aP(x,t)/P_b}{1+a}. \tag{4.31}$$

The key point of this equation is that the liquid flux through the capillary wall varies along the capillary. The reason for this is that a decrease in pressure along the capillary hampers the infiltration of the fluid through the capillary wall. This fact was taken into account in the simplest manner, by representing the pressure drop using a linear function, which equals 1 when $x = L_{cap}(t)$, with a single free parameter a.

The boundary conditions for Eqs. (4.30) and (4.31) are

$$P(x,t)\big|_{x=\pm L_{cap}(t)} = P_{at} - \frac{2\gamma}{R_{cap}} = P_b,$$

$$V(x,t)\big|_{x=\pm L_{cap}(t)} = \pm\frac{\partial L_{cap}(t)}{\partial t}, \tag{4.32}$$

$$L_{cap}(t=0) = L_0,$$

where $P_{at} = 10^5$ N/m^2 is the atmospheric pressure and γ is the surface tension coefficient.

Assuming $\gamma \propto 50 \times 10^{-3}$ N/m and $R_{\text{cap}} \propto 2 \times 10^{-5}$ m,

$2\gamma/R_{\text{cap}} \propto 0.1 \times 10^5$ N/m^2 ($\sim 0.1P_{\text{at}}$).

Thus, we can conclude that the influence of the surface tension on the process' kinetics is negligible. However, in the case of radial deformation of the capillary (e.g., due to bending), the influence of surface tension can become significant and result in a decrease in the displacement rate of the meniscus. Such phenomena were observed experimentally (see Fig. 4.24).

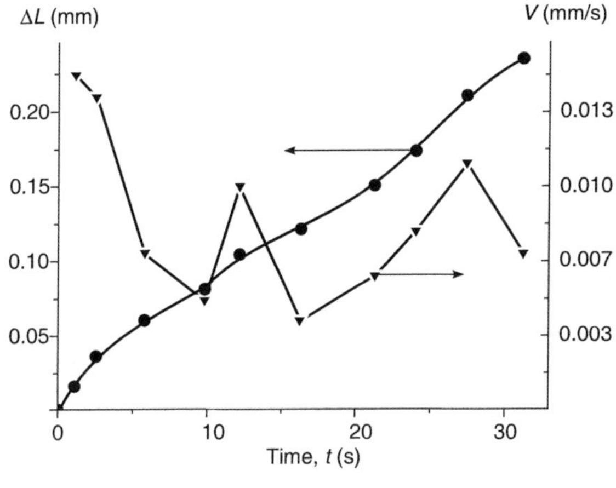

Figure 4.24 The meniscus displacement, $\Delta L(t) = L_0 - L_{\text{cap}}$ (t) (left axis) and its rate (right axis) versus time in the case of deformed capillaries. Reprinted from Ref. [65], Copyright 2007, with permission by the American Physical Society (https://doi.org/10.1103/PhysRevE.76.056303).

Equations (4.30) and (4.31), together with the boundary conditions (4.32), allow one to more fully investigate the evaporation process and the fiber's evolution. Also, the velocity distribution along the fiber allows one to analyze the displacement rate of the menisci being observed experimentally. Correlating the experimental observations with the aforementioned model provides with the actual physical parameters of the fibers.

4.5.4 Pressure and Velocity Distribution

The solution to Eqs. (4.30) and (4.31) can be obtained as follows: After the differentiation of Eq. (4.31) and substitution of fP/fx in Eq. (4.30), we obtain

$$\xi^2 \frac{\partial^2 V(x,t)}{\partial x^2} - V(x,t) = 0, \tag{4.33}$$

and the velocity is

$$V(x,t) = -A(t)\sinh(x/\xi), \tag{4.34}$$

where $\xi = \frac{1}{4}\sqrt{(1+a)R_{\text{cap}}^3 d_{\text{w}} P_{\text{b}} / aD_{\text{s}}\eta}$.

The pressure $P(x, t)$ can be found using Eq. (4.30) and is expressed as

$$P(x,t) = \frac{1}{a}P_{\text{b}}\left[-1 + (1+a)(\tau/\xi)A(t)\cosh(x/\xi)\right], \tag{4.35}$$

where $\tau = R_{\text{cap}} d_{\text{w}}/2D_{\text{s}}$.

Using the first boundary conditions (4.32), we find that

$$A(t) = \frac{\xi}{\tau\cosh\left(L_{\text{cap}}(t)/\xi\right)}. \tag{4.36}$$

Thus, we find that the velocity distribution of liquid flux along a capillary is

$$V(x,t) = -\frac{\xi\sinh(x/\xi)}{\tau\cosh\left(L_{\text{cap}}(t)/\xi\right)}, \tag{4.37}$$

and the pressure drop along a capillary is

$$\Delta P(x,t) = P_{\text{at}} - P(x,t) \tag{4.38}$$

$$= \frac{1+a}{a}P_{\text{b}}\left[1 + \varepsilon - \frac{\cosh(x/\xi)}{\cosh\left(L_{\text{cap}}(t)/\xi\right)}\right],$$

where $\varepsilon = \frac{a}{1+a}\frac{2\gamma}{R_{\text{cap}}}P_{\text{b}} \le 0.05$.

4.5.5 Meniscus Moving (Kinetics of Solvent Evaporation)

In order to determine the displacement rate of the meniscus, the second boundary condition (4.32) is used

$$\frac{\partial L_{\text{cap}}(t)}{\partial t} = -\frac{\xi}{\tau}\frac{\sinh\left(L_{\text{cap}}(t)/\xi\right)}{\cosh\left(L_{\text{cap}}(t)/\xi\right)}. \tag{4.39}$$

The solution to Eq. (4.39) is

$$\sinh (L_{cap}(t)/\xi) = \sinh (L_0/\xi) \cdot \exp(-t/\tau) \tag{4.40}$$

At the initial stage, when $L_{cap}(t)/\xi >>1$, Eq. (4.40) can be transformed into

$$\exp(L_{cap}(t)/\xi) = \exp(L_0/\xi) \cdot \exp(-t/\tau). \tag{4.41}$$

Hence, the meniscus moves at a constant rate, $-\xi/\tau$,

$$L_0 - L_{cap}(t) \approx \xi t/\tau. \tag{4.42}$$

For the opposite case, when $t/\tau >>1$, i.e., $L_{cap}(t)/\xi << 1$, Eq. (4.40) is

$$L_{cap}(t) = \xi \sinh(L_0/\xi) \,.\, \exp(-t/\tau). \tag{4.43}$$

Hence, the meniscus displacement rate decreases exponentially,

$$V_m(t) = \frac{\partial L_{cap}(t)}{\partial t} \approx -\frac{\xi}{\tau}\sinh(L_0/\xi)\cdot\exp(-t/\tau). \tag{4.44}$$

These solutions reflect the asymptotic behavior presented in Fig. 4.22.

Fitting the two parameters (ξ and τ) of Eq. (4.40) to typical experimental data results in $\xi \approx 0.15 \times 10^{-3}$m and $\tau \approx 1.5$ sec (see Fig. 4.22). At the same time, the values of these parameters can be determined based on their definition (see Eqs. 4.34 and 4.35). Assuming that $P \approx 10^5$ *Pa*, $D_s \propto 5 \times 10^{-13}$ m^2/sec and $\eta \propto 100$ *Pa* . sec $\eta \propto 100$ *Pa* $\cdot$ sec, we noticed that for capillaries with a radius of about $R_{cap} \approx 10^{-5}$ m and a wall thickness of about $d_w \approx 0.5 \times 10^{-6}$ m, the aforementioned parameters of the model are $\xi \approx 0.35 \times 10^{-3}$ m and $\tau \approx 5$ sec. Thus, the two independent methods of parameter determination result in similar values, which shows a good agreement between the suggested theoretical model and the experimental data.

Thus, this analysis has demonstrated that in the first few seconds after fabrication, some amount of residual solvent is encapsulated inside the electrospun nanofibers. This solvent plays the role of a plasticizer, allowing the partial relaxation of stretched macromolecules.

4.5.6 Experimental Proof of Post-fabrication Relaxation

This outcome can be verified with the help of the following experiment. During the spinning of any polymer nanofibers, a portion of these fibers should be collected in a bath with liquid nitrogen. These electrospun nanofibers will be frozen, and no relaxation will be possible. Before heating the frozen nanofibers up to room temperature, they should be kept at a low pressure in order to sublimate the residual solvent. Thereafter, the behavior of these nanofibers should be compared to that of the fibers, collected in ambient atmosphere, without the aforementioned procedure suppressing the relaxation. Such experiment was fulfilled in the "Nano engineering group" Technion by the graduate student M. Kaplan, in the framework of his training (unfortunately, this study was not continued due to difficulties in the collection of well-aligned mat in liquid nitrogen).

First of all, the difference in heterogeneity between frozen (unrelaxed) and relaxed electrospun nanofibers can be seen visually (see Fig. 4.25). Indeed, in spite of the fact that both groups of fibers (frozen and relaxed ones) demonstrate high porosity, the heterogeneity of the frozen structure is much higher compared to the heterogeneity in the relaxed fibers (compare Figs. 4.25a and 4.25b).

Figure 4.25 The difference in heterogeneity between frozen (unrelaxed) (a) and relaxed (b) electrospun nanofibers.

The mechanical testing also demonstrates the difference between these two structures: The elastic modulus of the fibers collected in liquid nitrogen (i.e., these fibers are frozen and cannot relax) is much higher (more than twice) compared to the elastic modulus of the fibers collected in ambient atmosphere at room temperature (i.e., a partial relaxation occurs in these fibers) (see Fig. 4.26, where the tensile tests for the frozen (unrelaxed) and relaxed electrospun nanofibers are shown as the strain–stress dependencies).

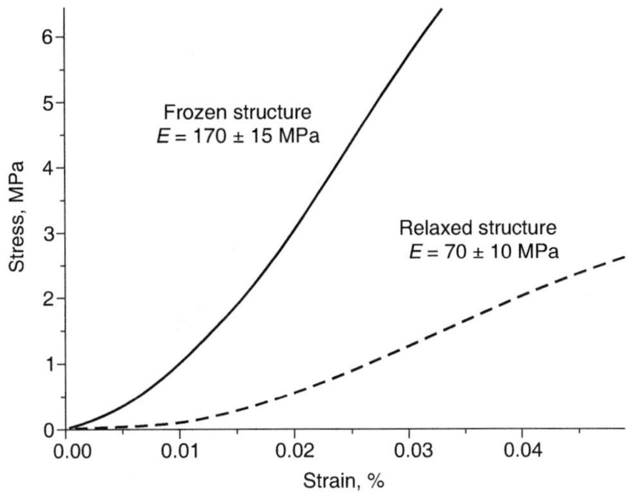

Figure 4.26 Tensile test for the frozen (unrelaxed) and relaxed electrospun nanofibers.

4.6 Post-processes in Electrospun Tubular Nanofibers

In order to analyze the possible post-processes in electrospun tubular nanofibers, we have to determine the pressure distribution along the fiber, which provides information about the elastic energy being stored in the fibers during evaporation as a result of their radial compression. This elastic energy accounts for additional phenomena observed in the experiments, for example, the formation of gas bubbles in the fibers or modifications in the fiber structure, such as buckling.

It would appear that once the elastic energy of the fiber system increases, it tends to find another state in order to lower its energy.

However, kinetic barriers can substantially influence the choices available to the system. Therefore, it is essential to explore the various possibilities that would allow the system energy to decrease.

The pressure drop across the capillary shell gives rise to a tangential stress, which varies along the capillary and depends on the menisci displacement (see Fig. 4.27)

$$\sigma_\theta = (R_{cap}/d_w)\Delta P. \tag{4.45}$$

This stress produces an elastic strain on the capillary shell, which can result in additional post-processes minimizing the elastic energy of the fiber shell.

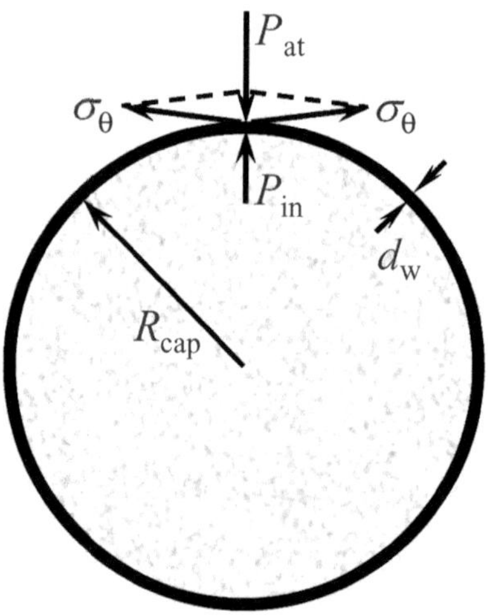

Figure 4.27 The scheme of tangential stress created by the differences between the inside (P_{in}) and outside (P_{at}) pressures of a capillary, resulting in elastic compression of the capillary shell. Reprinted from Ref. [65], Copyright 2007, with permission by the American Physical Society (https://doi.org/10.1103/PhysRevE.76.056303).

4.6.1 Formation of Gas Bubbles

One possible means of decreasing the elastic energy of a fiber is to replace a single long slug with several smaller ones separated by areas filled by both solvent vapor and air that has penetrated through the shell and into the fiber, i.e., the formation of gas bubbles. In order to establish the conditions for bubble formation, the elastic energy of the deformed fiber shell should be determined.

The density of the elastic energy, which depends on the location along the capillary and time, is $\Delta P(x, t)^2/2E$, where E is Young's modulus of the capillary shell under compression. Thus, the time-dependent elastic energy of a compressed capillary as a function of the length of the solvent slug is expressed as

$$W_{\mathrm{el}}(t) = W_{\mathrm{el}}^{(0)} \left\{ \frac{L_{\mathrm{cap}}(t)}{\xi} \left[1 + \frac{1}{2\cosh^2\left(L_{\mathrm{cap}}(t)/\xi\right)} \right] - \frac{3}{2}\tanh\left(\frac{L_{\mathrm{cap}}(t)}{\xi} \right) \right\}, \tag{4.46}$$

where $W_{\mathrm{el}}^{(0)} = \left(1 + 1/a\right)^2 2\pi R_{\mathrm{cap}}^3 P_{\mathrm{b}}^2 \xi / E d_{\mathrm{w}}$ and $L_{\mathrm{cap}}(t)$ is determined by Eq. (4.40).

In the case of a long slug splitting into two equal separated slugs, the elastic energy decreases and the energy gain can be represented as follows

$$\begin{aligned} \Delta W_{\mathrm{el}}(L) &= W_{\mathrm{el}}(L) - 2W_{\mathrm{el}}(L/2) \\ &= W_{\mathrm{el}}^{(0)} \left\{ 3\left[\tanh\left(\frac{L}{2\xi} \right) - \frac{1}{2}\tanh\left(\frac{L}{\xi} \right) \right] \right. \\ &\quad \left. + \frac{L}{2\xi}\left[\cosh^{-2}\left(\frac{L}{\xi} \right) - \cosh^{-2}\left(\frac{L}{2\xi} \right) \right] \right\}. \end{aligned} \tag{4.47}$$

If this gain of energy exceeds the energy required to create the two new menisci, $2\pi R_{\mathrm{cap}}^2 \gamma$ (here $\gamma \propto 50 \times 10^{-3}$ N/m is the surface tension coefficient), bubble formation takes place.

The function $\Delta W_{\mathrm{el}}(L)$ tends to a constant value 3/2 as $f(x) \approx 1.5 - 2x^2/1.5 - 2x\exp(-x)$, if $x >> 1$, and tends to zero as $f(x) \approx 2x^2/15$, if $x << 1$ (see Fig. 4.28).

In order to find the threshold where bubble formation can occur (i.e., a slug splits into two slugs), the function $f(x)$ can be compared with the dimensionless ratio of the surface tension energy $2\pi R_{\mathrm{cap}}^2 \gamma / W_{\mathrm{el}}^{(0)} = \gamma E d_{\mathrm{w}} / R_{\mathrm{cap}} P_{\mathrm{b}}^2 \xi \propto 0.2$ ($E \sim 0.5 GPa$). Thus, bubble formation is possible if the length of the slug exceeds $L_{\mathrm{cap}}^{(\mathrm{cr})} \approx 1.5\xi \propto 0.3 \times 10^{-3}\,\mathrm{m}$.

This fact was confirmed by the experimental observation that within a few seconds after the fabrication of a tubular nanofiber, a large slug splits into several smaller slugs separated by gas bubbles.

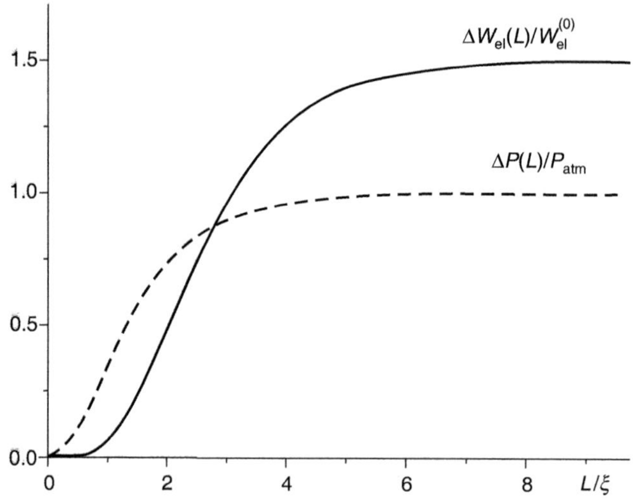

Figure 4.28 The reduced gain of energy $\Delta W_{el}(L)/W_{el}^{(0)}$ (solid line) and the reduced pressure difference in the center of a column of liquid $\Delta P(L)/P_{atm}$ (dashed line) versus reduced column length L/ξ. Reprinted from Ref. [65], Copyright 2007, with permission by the American Physical Society (https://doi.org/10.1103/PhysRevE.76.056303).

4.6.2 Buckling of Tubular Nanofiber

However, the more interesting phenomenon is the buckling of tubular nanofiber, which can occur on the final stage of solvent evaporation [66]. This phenomenon is related to the fact that once evaporation through the shell begins, a pressure drop across the fiber shell develops, whereby the external pressure becomes higher than the internal pressure, and the tubular fiber is able to deform. It is a well-known fact that for hollow cylindrical surfaces, such shape modifications are negligible if the pressure drop is relatively small; however, if this pressure drop exceeds some critical value, the cylinder becomes unstable and its shape can vary dramatically. In this case, the tubular fiber loses its structural stability and fiber buckling occurs. Taking into account the geometrical and mechanical properties of a capillary, the critical pressure is shown to be [67]

$$\Delta P_{cr} = \frac{Ed_w^3}{12R_{cap}^3} \cdot \frac{n^2-1}{1-\nu^2} \propto 1.3 \times 10^2 \cdot \frac{n^2-1}{1-\nu^2}\, Pa, \tag{4.48}$$

where v is Poisson's ratio, which can be taken equal to 0.5 and n = 2, 3,... is the azimuthal wave number of the buckling mode. Thus, if the pressure drop ΔP exceeds 6×10^2 *Pa*, buckling of the capillary occurs. Therefore, based on Eq. (4.38), and assuming that the maximal pressure drop is at the center of the slug, we calculate the critical length for buckling to occur when $L_{cap} > 1.1 \times 10^{-2}\ \xi \propto 1.1 \times 10^{-6}$ m ($2L_{cap}$ is the length of the slug). Note that this length is less than the typical radius of the microtube. Consequently, the conditions that promote structural instability exist during the entire solvent evaporation process. It would seem that the conditions for fiber buckling are most favorable during the initial stage of the evaporation process when the fibers are filled with solvent and the slugs are relatively long. However, fiber buckling was observed only during the final stage of evaporation. Fortunately, there is a simple explanation for this behavior.

During the initial stage of the process, the tubular fibers are filled with solvent, which prevents the fiber from buckling. Since any small deformation of a fiber would result in a decrease in the internal volume, any buckling/deformation could occur only after a portion of the solvent's volume has been removed through evaporation. This would require a flux having a direction opposite to the initial one. But this flux directional change can only occur when there is an increase in local pressure. In a region of fiber deformation, the difference between the external and internal pressures sharply decreases; therefore, no further deformation or buckling of the fiber can take place.

Buckling in a filled tubular fiber is not possible when the column slug is long and the meniscus does not influence the pressure inside the fiber. However, if the slug is short enough, the surface tension forces can disturb the aforementioned stability mechanism and fiber buckling can occur. Indeed, the forces of surface tension decrease the pressure of the solvent in the meniscus region ($P_m = P_{atm} - 2\gamma/R_{cap}$), so as to maintain the critical difference between external and internal pressures in a given zone of fiber deformation ($\Delta P_{cr} = P_{atm} - P_{in}$, ΔP_{cr}, is given in Eq. (4.48)). At the same time, the pressure decrease provides the necessary pressure gradient along the fiber ($\Delta P_{cap} = (P_{in} - P_m)/L_{cr}$) for flux to be created by the removal of solvent with a rate exceeding that of meniscus movement (see Fig. 4.29). Thus, fiber buckling will take place if the following condition exists

$$P_{\text{in}} = P_{\text{atm}} - \frac{Ed_{\text{w}}^3}{12R_{\text{cap}}^3} \cdot \frac{n^2 - 1}{1 - \nu^2} > P_{\text{m}} = P_{\text{atm}} - \frac{2\gamma}{R_{\text{cap}}} \tag{4.49}$$

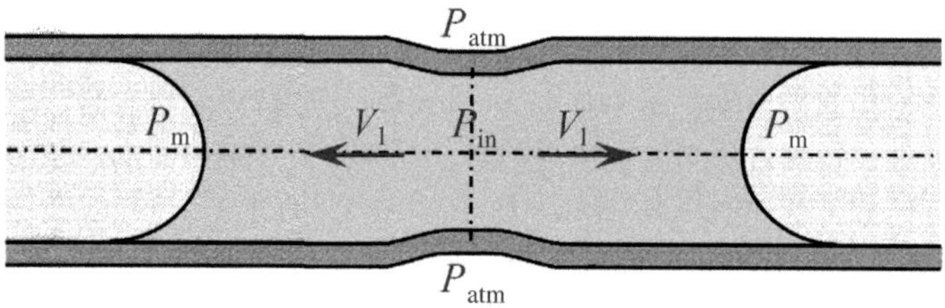

Figure 4.29 A scheme of fiber buckling in the case of relatively short slugs. Due to the pressure difference $P_{\text{in}} = P_{\text{atm}} - \frac{n^2-1}{1-\nu^2} Ed_{\text{w}}^3 / 12R_{\text{cap}}^3 > P_{\text{m}} = P_{\text{atm}} - 2\gamma / R_{\text{cap}}$, the solvent flux moves away from the buckling region. Reproduced with permission from Ref. [66], Copyright 2008, IOP Publishing. All rights reserved (https://doi.org/10.1088/0022-3727/42/1/015507).

This equation demonstrates that the phenomenon of fiber buckling can take place only during the final stage of evaporation when the slugs are sufficiently short. Assuming Poiseuille's flow

$$V_1 = \frac{R_{\text{cap}}^2}{8\eta} \frac{\Delta P_{\text{cap}}}{L_{\text{cr}}} = \frac{R_{\text{cap}}^2}{8\eta} \frac{P_{\text{in}} - P_{\text{m}}}{L_{\text{cr}}} = \tag{4.50}$$

$$= \frac{1}{L_{\text{cr}}} \frac{R_{\text{cap}}^2}{8\eta} \left(\frac{2\gamma}{R_{\text{cap}}} - \frac{Ed_{\text{w}}^3}{12R_{\text{cap}}^3} \cdot \frac{n^2 - 1}{1 - \nu^2} \right) \propto \frac{\xi}{\tau}$$

The critical length of the solvent slug is calculated as

$$L_{\text{cr}} = \frac{\tau}{\xi} \frac{R_{\text{cap}}^2}{8\eta} \left(\frac{2\gamma}{R_{\text{cap}}} - \frac{Ed_{\text{w}}^3}{12R_{\text{cap}}^3} \cdot \frac{n^2 - 1}{1 - \nu^2} \right) \propto 0.07 \times 10^{-3} \text{ m} \tag{4.51}$$

This scenario was confirmed in the experimental study. During the initial stage of the evaporation process, no fiber buckling was observed. However, as the solvent evaporation progressed, regions containing no solvent began to constitute a noticeable part of the fiber. As these regions offered no resistance to fiber buckling, radial buckling was able to occur and, indeed, was observed. The typical length of these solvent slugs was 100 ÷ 200 μm or less, while the critical length as obtained in (4.51) was $2L_{cr} \propto 140$ μm. Typical optical images of the buckled zones along a fiber, and the cross section of a collapsed fiber after evaporation are presented in Fig. 4.30a and Fig. 4.30b, respectively.

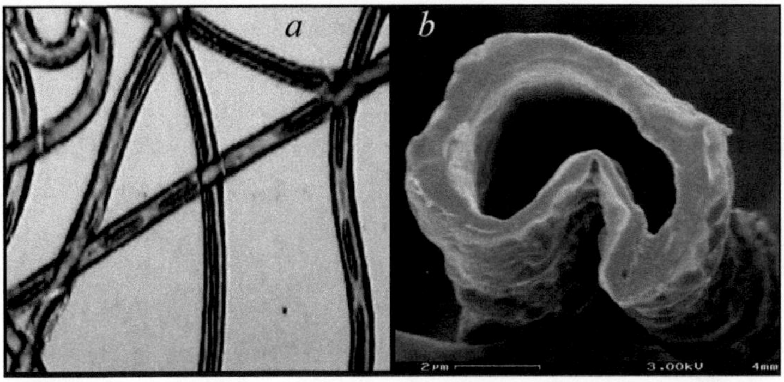

Figure 4.30 Typical optical images of the buckled zones along a fiber (a) and the cross section of a collapsed fiber after evaporation (b). Reproduced with permission from Ref. [66], Copyright 2008, IOP Publishing. All rights reserved (https://doi.org/10.1088/0022-3727/42/1/015507).

At first glance, it would seem that the buckling phenomenon has to occur whenever the system parameters result in the square brackets in Eq. (4.28) being positive, i.e., when the surface tension γ is high enough, the fiber radius R_{cap} small enough, and the polymer's Young's modulus E not too high. However, along with systems consisting of collapsed nanofibers or of non-collapsed ones, one can get systems with only partially collapsed nanotubes. The cause is that the initial stage of fiber buckling (when structural instability first develops) is of a fluctuating nature. Therefore, its realization takes some characteristic time t_b (time to buckling). On the other hand, further solvent evaporation results in the decrease in the slug length being accompanied by a decreased pressure difference. As a result, after some time Δt, the slug length decreases below the value L_{min}, at which point the pressure difference becomes lower than the critical one and no buckling can occur in the system.

The slug length L_{min} can be determined by equating the right-hand sides of Eqs. (4.38) and (4.48):

$$L_{min} \simeq \sqrt{\frac{R_{cap}^3 d_w P_{atm}}{8\mu D_s}} \operatorname{arccosh}\left(\frac{1}{1 - Ed_w^3 / 3R_{cap}^3 P_m}\right) \tag{4.52}$$

Using the fact that in the final stage of solvent evaporation the slug length decreases exponentially (see 4.43), the time interval Δt can be calculated as

$$\Delta t = \frac{R_{\text{cap}} d_{\text{w}}}{2D_{\text{s}}} \ln\left(\frac{L_{\text{cr}}}{L_{\text{min}}}\right) \tag{4.53}$$

Note that the ratio $L_{\text{cr}}/L_{\text{min}}$ depends on the fiber's parameters (fiber radius R_{cap}, fiber thickness d_{w}, and its modulus E), core solvent surface tension γ, and the external pressure P_{atm}. However, it does not depend on the diffusion coefficient D_{s} or the core solution viscosity μ.

Since the fiber collapse can be considered a Poisson process at which at least one arrival occurs, the buckling probability is exponentially dependent on the time interval Δt

$$w(\Delta t) = 1 - \exp(-\Delta t/t_{\text{b}}) \tag{4.54}$$

For large Δt ($\Delta t >> t_{\text{b}}$), the probability w tends to 1, i.e., we get collapsed nanofibers. In the opposite case for small Δt ($\Delta t << t_{\text{b}}$), the probability w tends to 0, and no collapse takes place in the system. In the case of intermediate values of Δt, that is $\Delta t \propto t_{\text{b}}$, we can observe both collapsed and non-collapsed nanofibers. Therefore, the time interval Δt can be an effective means for predicting the final form (tubular or ribbon-like) of the produced nanofibers.

Unfortunately, there is no a-priori information on the time to buckling, t_{b}, and theoretical estimation of this parameter is a very complicated problem, which is hard to solve. For this reason, the value of t_{b} can be estimated based only on experimental observation, and these estimations are presented in Fig. 4.31.

Figure 4.31 also shows the results of fitting of the experimental data with the help of exponential distribution function (4.54) containing one free parameter—time to buckling, t_{b}. It turns out that the time to buckling is about 0.8 sec ($t_{\text{b}} = 0.799$ sec), which is in good agreement with the visual observations.

The above post-processes in electrospun tubular nanofibers, which are observed experimentally (formation of gas bubbles, buckling of tubular nanofiber, and, maybe, others), are good validations for our assumption regarding a residual solvent, encapsulated inside electrospun nanofibers immediately after fabrication; and this residual solvent provides a partial relaxation of the stretched polymer matrix of electrospun nanofibers.

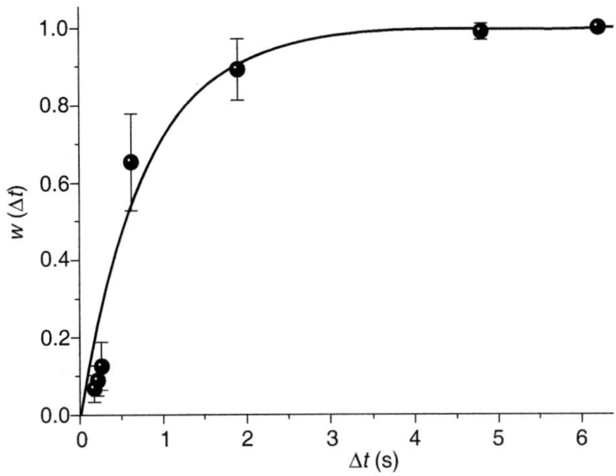

Figure 4.31 Fitting of the experimental data using exponential distribution function, $w(\Delta t) = 1 - \exp(-\Delta t/t_b)$. The time to buckling is $t_b = 0.799$ s. Reproduced with permission from Ref. [66], Copyright 2008, IOP Publishing. All rights reserved (https://doi.org/10.1088/0022-3727/42/1/015507).

Chapter 5

Polymer Dynamics in Semi-dilute Solution During Electrospinning

A progress in understanding the features that influence the behavior of electrospun polymer nanofibers requires examination of their internal structure, more specifically the supermolecular structures in the amorphous regions of the polymer matrix, as well as crystallite ordering. At the initial stages of such analyses, the impact of fabrication conditions upon the final state of the polymer matrix of as-spun nanofibers must be determined. In doing so, the high strain rate (of the order 10^3 s^{-1}) acting during electrospinning [40–42, 45, 52, 58, 59] must be considered. This dominant factor is believed to cause stretching and orientation of polymer chains, as indicated by in-process measurements of jets, using birefringence [68] and Raman [69] techniques.

In parallel, extremely rapid solvent evaporation, discussed earlier (see Section 4.4 "Solvent evaporation and polymer solidification"), adversely affects the polymer matrix macro-structure of as-spun nanofibers. Indeed, rapid solvent evaporation first leads to the formation of a solid skin, followed by further evaporation from the liquid core, leaving voids previously occupied by solvents and allowing partial relaxation of the matrix. This frequently induces generation of a heterogeneous and porous fiber structure [70, 71]. Theoretical analyses [61, 72], confirmed by experimental

Electrospun Polymer Nanofibers
Arkadii Arinstein

ISBN 978-981-4745-27-7 (Hardcover), 978-1-315-36462-9 (eBook)
www.panstanford.com

observations [70] and computerized simulations [60], demonstrated that a sharp increase in polymer density at the fiber/vapor interface is induced upon rapid solvent evaporation, consequently increasing heterogeneity and porosity. Thus, heterogeneous, high-porosity fiber structures can be formed under certain electrospinning conditions (see Fig. 5.1a).

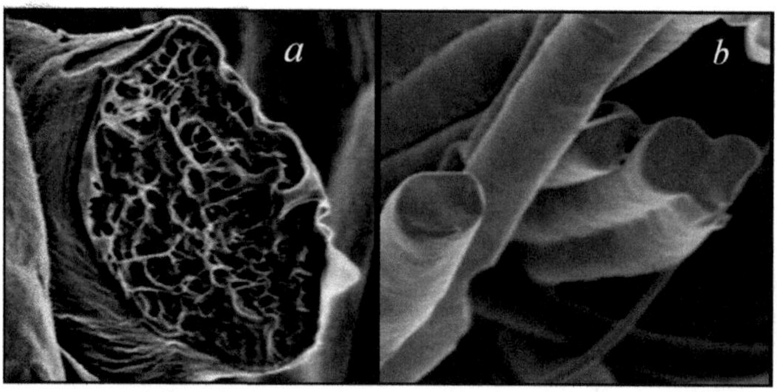

Figure 5.1 SEM images of electrospun nanofibers fabricated from 10 wt% PCL (M_w = 80 kDa) in DCM/DMF (75 : 25 wt%), in an electric field of 0.63 kV/cm . Heterogeneous fibers fabricated at a flow rate of 20 ml/h (a), and homogeneous fibers fabricated at a flow rate of 3 ml/h (b). Reprinted from Ref. [73], Copyright 2011, with permission from John Wiley and Sons.

At the same time, skin formation prevents further stretching and orientation of polymer chains within the semi-liquid fiber core. Relaxation of the nonequilibrium state of the stretched macromolecules consequentially occurs, with no detectable influence of the fabrication conditions on the final state of the polymer matrix of as-spun nanofibers. On the other hand, spinning conditions resulting in homogeneous as-spun fiber structures can be selected (see Fig. 5.1b). In such cases, relaxation of stretched polymer chains is suppressed, and the effect of fabrication conditions on the final state of the nanofiber polymer matrix should be noticeable.

While the physical mechanism resulting in the formation of the porous structure of as-spun nanofibers is well understood [61, 70, 72], the physical principles governing the generation of the homogeneous fiber structure demand clarification. In order to identify the required mechanism allowing to fabricate homogeneous fibers, the evolution of the polymer system within the semi-dilute solution jet during electrospinning should be analyzed. The key points of such analysis are the following two assumptions: polymer system in a spinnable solution is an interconnected topological network of subchains, and during electrospinning, this network undergoes substantial stretching [74, 75].

5.1 Theoretical Model Describing Stretching of a Polymer Topological Network

5.1.1 Initial State of Entangled Semi-dilute Polymer Solution

As was mentioned earlier, spinnable polymer solutions are semi-dilute, highly entangled. Therefore, the polymer system is assumed to be a network whose connectivity is provided by topological knots [52]. The sections of macromolecules between two adjacent topological knots are called subchains, which show a conformation at equilibrium that corresponds to their nonzero end-to-end distance, even in the absence of external force. This end-to-end distance is equal to the distance between two topological knots or to an average mesh size, ξ_0, of the network in a semi-dilute solution

$$\xi_0 \propto R_0(c/c^*)^{\nu/(1-3\nu)}, \tag{5.1}$$

where $R_0 \propto aN^{\nu}$ is the end-to-end distance of a polymer coil; c is the polymer concentration in the solution; c^* is the crossover concentration of macromolecule overlap; N is the number of monomers in a polymer chain (polymerization degree); and a is the monomer scale. Solvents usually used in such systems are good solvents (close to θ-condition), so it is reasonable to assume that $\nu \cong 0.5$. Assuming Gaussian statistics of subchains, one can estimate the number of monomers in a subchain as follows:

$$N_s \cong (\xi_0/a)^2 \cong N(c/c^*)^{-2} \tag{5.2}$$

Thus, the elements of the system in question are the above-defined subchains, which form the polymeric physical network with topological links arising as a result of a chain entanglement. This topological network provides the required viscoelasticity of the polymer solution. Moreover, under rapid deformation, this physical network can be regarded as a gel (the high-elastic state) and this fact is of high importance for further analysis.

5.1.2 Axial Stretching of an Entangled Polymer Network during Electrospinning

The polymer network in question can be approximated by a lattice model of "beads" and linear "springs," similar to the Rouse model. Each bead represents a topological knot and is connected to six adjacent beads by springs, or polymer subchains, demonstrating Gaussian statistics, leading to a linear force-elongation relationship (see Fig. 5.2a). The averaging of the system over the jet cross section results in a 1D chain of springs that interconnects the beads, each having an effective subchain mass m. The beads are influenced by an effective hydrodynamic force proportional to the effective subchain size, $a_{\mathrm{eff}} \propto \xi_0$, as well as entropic elastic forces from its two neighbors (see Figs. 5.2b and 5.2c).

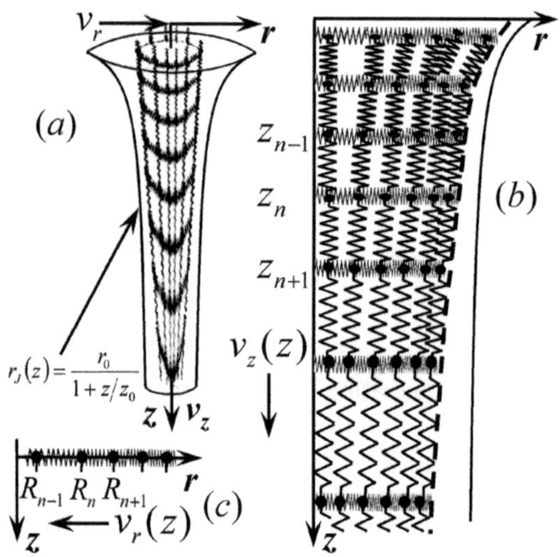

Figure 5.2 Illustration of polymer network stretching in an electrospinning jet (a). Definition of an effective 1D system describing polymer network stretching in the axial direction (b). Definition of an effective 1D system describing polymer network contraction in the radial direction (c). Reprinted from Ref. [74], Copyright 2011, with permission by the American Physical Society (https://doi.org/10.1103/PhysRevE.84.041806).

The dynamics of this chain of springs can be described by the following difference-differential equation:

$$m\frac{d^2 z_n}{dt^2} = a_{\mathrm{eff}}\eta\left[v_z(z_n) - \frac{dz_n}{dt}\right] +$$

$$+\frac{T}{\xi_0^2}\{[z_{n+1}-z_n-\xi_0]-[z_n-z_{n-1}-\xi_0]\}, \tag{5.3}$$

where η is the effective viscosity of the dilute solution surrounding the polymer network; v_z is the jet axial velocity from Eq. (4.6); T/ξ_0^2 is the entropic elasticity of polymer subchains, where T is the temperature in units of the Boltzmann constant k_B.

The difference equation (5.3) can be approximated by the following differential equation:

$$\frac{d^2z(l,t)}{dt^2}=\frac{a_{\text{eff}}\eta}{m}\left[v_z(z)-\frac{dz(l,t)}{dt}\right]+c_{\text{net}}^2\frac{d^2z(l,t)}{dl^2}, \tag{5.4}$$

where $c_{\text{net}}=\sqrt{T/m}\sim 0.3$ m/sec is the "sound" velocity in the polymer network and $l=\xi_0 n$. As the polymer network demonstrates only entropic elasticity of a semi-dilute solution, it is much lower than the sound velocity in solid polymer systems. The polymer is at equilibrium at the cross section $z=-z_0$, while free boundary conditions are assumed at $z=+\infty$.

The partial differential equation (5.4) allows one to find a self-similar solution. Indeed, upon introduction of the dimensionless, self-simulated variable $\varsigma=(l+v_0t)/z_0$ and the function $\hat{z}(\varsigma)=z(l,t)/z_0$, Eq. (5.4) can be rewritten as an ordinary differential equation:

$$\frac{d^2\hat{z}(\varsigma)}{d\varsigma^2}=-\alpha\left[\hat{v}_Z(\hat{z})-\frac{d\hat{z}(\varsigma)}{d\varsigma}\right], \tag{5.5}$$

where $\alpha=z_0v_0a_{\text{eff}}\eta/\left[m\left(c_{\text{net}}^2-v_0^2\right)\right]\sim 10^3\div 10^4>>1$ is a dimensionless parameter and $\hat{v}_Z(\hat{z})=v_z(z_0\hat{z})/v_0$.

Using the substitution $d\hat{z}(\varsigma)/d\varsigma=P(\hat{z})$, the second-order differential equation (5.5) can be reduced to a first-order equation

$$\frac{dP(\hat{z})}{d\hat{z}}=-\alpha\left[\frac{\hat{v}_Z(\hat{z})}{P(\hat{z})}-1\right] \tag{5.6}$$

The solution to this equation (5.6) can be obtained using $(1/\alpha)$ approximation. Assuming that

$$\frac{\hat{v}_Z(\hat{z})}{P(\hat{z})}-1=\frac{1}{\alpha}P_1(\hat{z})+\ldots \tag{5.7}$$

we obtain $P_1(\hat{z}) = -d\hat{v}_z(\hat{z})/d\hat{z}$, so within accuracy of $(1/\alpha)^2$, the function $P(\hat{z})$ is

$$P(\hat{z}) = \frac{d\hat{z}(\varsigma)}{d\varsigma} = \frac{dz(l,t)}{dl} \approx \frac{\hat{v}_z(\hat{z})}{1 - \frac{1}{\alpha} d\hat{v}_z(\hat{z})/d\hat{z}}, \tag{5.8}$$

and the function $\hat{z}(\varsigma)$ can be obtained as an inverse function

$$\varsigma(\hat{z}) = \int_0^{\hat{z}} \frac{dx}{\hat{v}_z(x)} - \frac{1}{\alpha}\ln\left(\hat{v}_z(\hat{z})\right) \tag{5.9}$$

The stretching of the polymer network is characterized by Eq. (5.8), which allows one to determine the stretched mesh size, $\xi_{||}$, along the jet

$$\xi_{||} = \xi_0 \frac{dz(l,t)}{dl} \approx \frac{\xi_0 \hat{v}_z(\hat{z})}{1 - \frac{1}{\alpha} d\hat{v}_z(\hat{z})/d\hat{z}} \tag{5.10}$$

The obtained dependence $\xi_{||}/aN_s$ is shown in Fig. 5.3 (dashed dot-dot line) with respect to the relative position, z/z_0, along the jet, for longitudinal velocity, $v_z \approx v_0(1 + z/z_0)^2$, from Eq. (4.6).

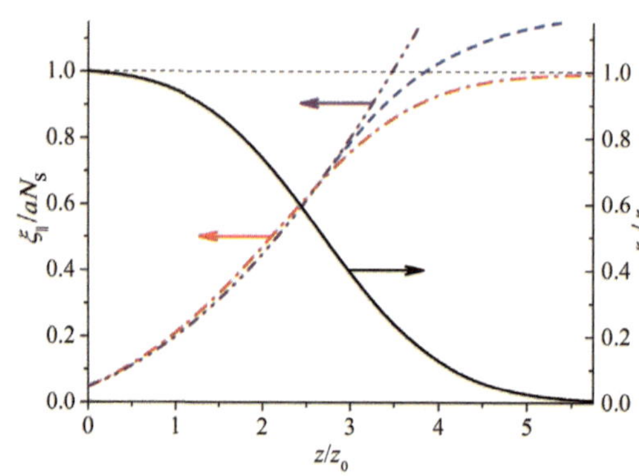

Figure 5.3 Polymer network conformation. Relative axial stretching, $\xi_{||}/aN_s$, (dashed dot-dot, dashed-dot and dashed lines), and relative radial contraction, $\xi_\perp/\xi_0$, (solid line) versus the relative axial position, z/z_0 (z_0 = 0.29). The results were obtained in the framework of the theoretical model (dashed dot-dot line) and by the numerical simulations (dashed-dot and solid lines). The physically reasonable behavior taking into account disentanglement is presented by the dashed line. Reprinted from Ref. [74], Copyright 2011, with permission by the American Physical Society (https://doi.org/10.1103/PhysRevE.84.041806).

Note that the stretching of the polymer subchains exceeds the maximal possible elongation corresponding to fully stretched subchains. Such type of dependence contradicts the physical meaning; therefore, Eq. (5.10) should be modified in such a manner that the maximum stretching of subchains would be restricted. A term that provides the unlimited increase in an effective modulus of an almost fully stretched subchains that, in turn, restricts their further stretching can be introduced into Eq. (5.3) by replacing the linear elasticity of polymer subchains, T / ξ_0^2, with the nonlinear $T / \left\{ \xi_0^2 \left[1 - \left(z_{n+1} - z_n \right) / aN_s \right] \right\}$. As a result, the stretching subchains will be restricted by their length. A similar behavior demonstrating the limit in the stretching of polymer subchains along the jet can be examined with the help of numerical simulations by the application of the 3D random walk model under external field (see Fig. 5.3, dashed-dot line). The details related to the numerical simulations in the framework of the 3D random walk model under external field can be found in [74]. However, such modifications do not account well for the system behavior and the additional processes associated with high levels of polymer subchain stretching. More specifically, almost fully stretched subchains begin to disentangle, thereby raising the subchain length ξ_0^2. In addition, the highly stretched polymer network begins to affect the effective viscosity of the solution, influencing the jet velocity. Furthermore, slipping of the solvent surface layer relative to the polymer network is also possible. Therefore, the correct dependence describing the length of the stretched mesh size, $\xi_{||}$, along the jet should be situated between two curves discussed earlier (see Fig. 5.3, dashed line).

For all these reasons, we would like to restrict ourselves to the simplest model describing only the initial stage of polymer network evolution inside the electrospinning jet. Although the proposed model does not describe the final state of the polymer matrix in electrospun nanofibers and is applicable only to the initial part of the jet where the stretching of a polymer system is not too high, it allows for analysis and understanding of the tendency in the evolution of the polymer during the electrospinning process.

The region where the stretched subchains are approaching full extension, i.e., $\xi_{||} \simeq aN_s$, can be estimated using Eq. (5.10) for $\alpha >> 1$

resulting in $\hat{v}_z(\hat{z}) \simeq N_s^{1/2}$. Using typical values for electrospinning solution jets of the experimental system (PEO 3 wt% in water, see [74]), one may estimate that the stretching occurs at $z_{\text{stretch}}/z_0 \simeq 3.5$.

Note that the transformation of subchains from a coil-like equilibrium state into a stretched state occurs as a continuous crossover, and no phase transition is observed, in contrast to the well-known coil-stretch transition, described by de Gennes [76]. Unlike stretching of an individual chain, during network stretching locally, the dominant force that provides this transformation is the elastic force, whereas the hydrodynamic forces give rise to the global stretching of the network. As a result, the network subchains are subjected to the action of the network portion situated farther along the jet, a force independent of local stretching. In contrast, the force acting on an individual polymer chain under an ultrahigh velocity gradient increases with macromolecule stretching. Similar behavior, continuous crossover from a coil-like state into a stretched state, was observed by Balabaev et al. upon examination of the state of an individual macromolecule under an external force acting on its ends [77].

5.1.3 Radial Contraction of an Entangled Polymer Network during Electrospinning

When analyzing the compression of the polymer network in a plane perpendicular to the jet, the process becomes stationary quite rapidly and can be presented as a 1D chain of springs (see Fig. 5.2c) in a steady state, described by the following differential equation:

$$\frac{T}{\xi_0^2}\left\{\left[R_{n+1}-R_n-\xi_\perp\right]-\left[R_m-R_{n-1}-\xi_\perp\right]\right\}+a_{\text{eff}}\eta v_r(R_n)=0, \quad (5.11)$$

which can be approximated by the following differential equation (using the velocity v_r from Eq. (4.6))

$$\frac{d^2R(\rho,z)}{d\rho^2}-\kappa^2(z)R(\rho,z)=0, \quad (5.12)$$

where $\rho = r/z_0$, $\kappa(z)=\kappa_0\left[\xi_0/\xi_\perp(z)\right]\sqrt{1+z/z_0}$, and $\kappa_0^2 = z_0 v_0 a_{\text{eff}}\eta/mc_{\text{net}}^2$ $=\left(v_0^2/c_{\text{net}}^2-1\right)\alpha$, ($\kappa_0 \sim 10 \div 10^2$).

The boundary conditions for Eq. (5.12) are

$$\frac{dR(\rho_0)}{d\rho} - z_0 + \frac{\xi_\perp(z)}{z_0}\kappa^2(z)R(\rho_0) = 0, \quad R(0) = 0, \tag{5.13}$$

where $\rho_0 \equiv \rho_0(z) = r_J(z)/z_0$ corresponds to the jet radius in the cross section z (see Eq. (4.7)), and $\xi_\perp$ is the lateral contraction of the mesh size due to the axial stretching $\xi_{||}$:

$$\xi_\perp = \frac{3\xi_0}{2 + \left[1 - \left(a\xi_{||} / \xi_0^2\right)^2\right]^{-1/2}} \tag{5.14}$$

The solution to Eq. (5.12) with the boundary conditions (5.13) is

$$R(\rho, z) = \frac{z_0 \sinh\left[\kappa(z)\rho\right] / \cosh\left[\kappa(z)\rho_0\right]}{\kappa(z)\left\{1 + \left[\xi_\perp(z) / z_0\right]\kappa(z)\tanh\left[\kappa(z)\rho_0\right]\right\}} \tag{5.15}$$

The polymer network radius $R_p(Z)$ is given at $\rho = \rho_0$ by the following form:

$$R_P(z) = \frac{z_0 \tanh\left[\kappa(z)\rho_0\right]}{\kappa(z)\left\{1 + \left[\xi_\perp(z) / z_0\right]\kappa(z)\tanh\left[\kappa(z)\rho_0\right]\right\}} \tag{5.16}$$

For small values of argument $\kappa(z)\,\rho_0 << 1$ and $\kappa(z)\xi_\perp(z)/z_0 << 1$, Eq. (5.16) demonstrates no radial contraction of the polymer network, yielding

$$R_p(z) = z_0\rho_0(z) = r_J(z) \tag{5.17}$$

For large values of the argument of "tanh" ($\kappa(z)\rho_0 >> 1$), one can assume that $\tanh[\kappa(z)\,\rho_0] \approx 1$, so that Eq. (5.16) can be simplified. Taking into account the fact that $\kappa(z)\,\xi_\perp(z)/z_0 << 1$, in case of finite z/z_0 one can write

$$R_P(z) = \frac{\xi_\perp(z)}{\xi_0}\frac{z_0}{\kappa_0\sqrt{1 + z / z_0}}, \tag{5.18}$$

and in the case of large z ($z/z_0 >> 1$), the polymer network has the following form:

$$R_P(z) = \frac{z_0^2}{\kappa_0^2\xi_0\left(1 + z / z_0\right)}, \tag{5.19}$$

indicating a constant ratio between the jet and polymer radii. However, the last asymptote lacks physical significance, as the effects dominant at this stage of the processes are not accounted for, thereby making the model inappropriate far from the jet start.

An approximation for the dependence of the polymer network radius $R_p(z)$

$$R_P(z) = \frac{\xi_\perp(z)}{\xi_0} r_J(z) \simeq \frac{3r_J(z)}{2 + \left[1 - \left(a\hat{v}_z(\hat{z})/\xi_0\right)^2\right]^{-1/2}}, \tag{5.20}$$

showing that the axial stretching is the dominant factor determining the network radius, while the radial hydrodynamic compression has a negligible effect. When presenting the relationship $R_p(z)/r_0$ as a function of position z along the jet and comparing it to Eq. (5.18), a rapid narrowing of the network radius is observed with respect to the jet radius (see Fig. 5.4). The expected outcome is a substantial rise in polymer concentration toward the jet center.

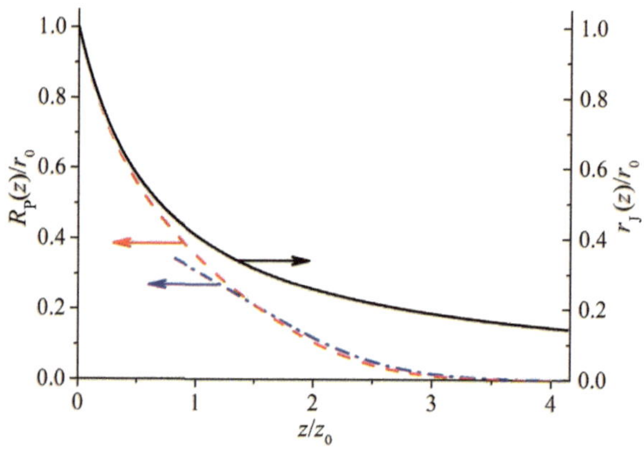

Figure 5.4 Polymer network contraction. Relative jet radius r_J/r_0 (solid line), and polymer network radius R_p/r_0: dashed line [Eq. (5.20)] and dashed-dot line [Eq. (5.18)], as a function of the relative axial position, z/z_0.

Summarizing the aforementioned results, one can conclude that no contraction of the polymer network occurs at the initial region of the jet. The intermediate asymptote demonstrates that the radial hydrodynamic effect is negligible, and that longitudinal stretching acts as the dominant factor affecting polymer network contraction, resulting in shortened distances between adjacent topological knots in the transversal direction. In the case of very large z ($z/z_0 >> 1$), the form of the polymer network conforms to that of the jet. Note that the last asymptote is barely discernible in a real system, due to rapid solvent evaporation in this region, which acts as the dominant factor determining the state of the polymer system.

5.2 Experimental Observations of Polymer System Evolution under High Stretching

Examining the structure of polymer solution inside electrospinning jets and specifically analysis of the evolution of the polymer-entangled network during electrospinning are very hard problems requiring modern complicated equipment. If the geometry and velocity of electrospinning jets can be studied by optical microscopy [58, 59], the tracing of the internal jet structure requires to utilize methods based on transmission spectroscopy, more specifically, on absorption of transmitted radiation.

5.2.1 Radial Contraction of an Entangled Polymer Network during Electrospinning

Taking into account the fact mentioned earlier that electrospinning is a very fast process accompanied by fast fluctuations in jet position, fast X-ray phase-contrast imaging [78] is suitable for our goal: analysis of the internal structure of polymer solution inside electrospinning jets. The point is that utilizing an extremely powerful X-ray beam generated from the synchrotron electron storage ring (32-ID Beamline, Advanced Photon Source, Argonne National Lab in Chicago) allows one to get necessary images within an extremely short exposure time (< 0.5 μs), eliminating the problems related to both high process rate and fast fluctuations in jet position.

A scheme of the experimental setup consisting of the X-ray imaging system and the electrospinning apparatus and a typical 16-bit grayscale image recorded with the help of this equipment are presented in Fig. 5.5 [74].

The measured intensity of transmitted X-ray beam is the result of the incident beam attenuation due to both absorption and scattering by the materials contained in the jet. Thus, the radiation transmission can be written using the Beer–Lambert law in the following form:

$$T(r,z) = g(r,z)\exp[-\alpha(r,z)d(r,z)], \tag{5.21}$$

where $g(r, z)$ is a form factor describing the scattering effect; $\alpha(r, z)$ is the absorption coefficient of polymer solutions; $d(r,z) = 2r_J(z)\sqrt{1-\left[r/r_J(z)\right]^2}$ is the length traveled by the beam through the jet at the radial position r; and $r_J(z)$ is the local jet radius at position z [79].

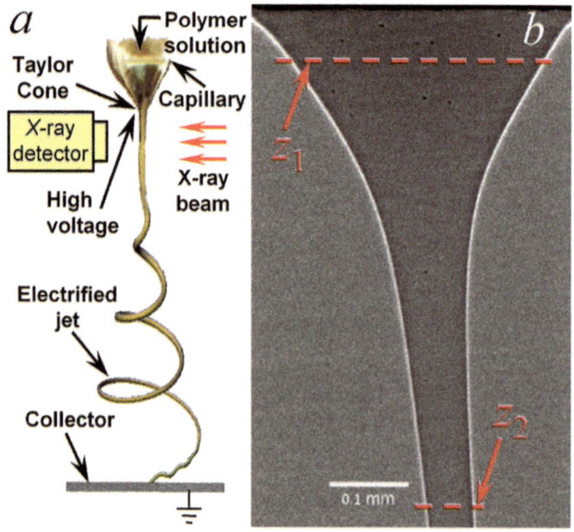

Figure 5.5 Schematics of electrospinning and fast in-process X-ray imaging. The imaged region is circled (a). Zooming on the Taylor cone. The lines at z_1 = 0.02 mm and z_2 = 0.02 mm indicate the cross sections of the absorption measurements across the electrospinning jet (b).

It is quite reasonable to assume that the absorption coefficient of polymer solutions can be calculated according to the mixing rule:

$$\alpha(r, z) = \varepsilon_P c_P\,(r, z) + \varepsilon_S c_S(r, z), \tag{5.22}$$

where ε_P and ε_S denote the X-ray mass absorption coefficients of the polymer and solvent, respectively, and $c_P(r, z)$ and $c_S(r, z)$ denote the mass concentrations.

In order to detect variations in polymer concentration across the jet, the experimentally measured intensity profile of the radiation transmission, $T_{\mathrm{exp}}(r, z)$, should be compared to the intensity calculated for the homogeneous polymer solution at rest, $T_{\mathrm{sim}}(r, z)$. (The last one can be found using a wave propagation simulation that took into account the overall optical effects, including both absorption and scattering.)

Assuming that both experiment and simulation result in the similar scattering effect (i.e., $g_{exp}(r, z)/g_{sim}(r, z) \cong 1$), normalization $T_{exp}(r, z)$ by $T_{sim}(r, z)$ reduces the influence of the scattering. Thus, the average variations in the measured absorption coefficient, $\Delta\alpha(r, z) = \alpha_{exp}(r, z) - \alpha_{sim}(r, z)$, with respect to the value of the fluid at rest, obtained by the wave propagation simulation can be calculated with the help of the following equation:

$$\Delta\alpha(r,z) = -\frac{1}{d(r,z)}\ln\left[\frac{T_{exp}(r,z)}{T_{sim}(r,z)}\right] \tag{5.23}$$

Although Eq. (5.23) cannot fully compensate for the scattering effects, the obtained $\Delta\alpha(r, z)$ is sufficient, as it represents a trend, rather than an exact value. Typical results of the change in the absorption coefficient across the jet are presented in the insets of Fig. 5.6. Almost no change in the absorption coefficient across the jet was observed close to the jet start, whereas a distinct change in the absorption coefficient was observed farther along the jet.

And, finally, required variations in the local polymer concentration can be calculated using the obtained variations in the absorption coefficient, $\Delta\alpha(r, z)$ from Eq. (5.23), and the normalization condition

$$\phi_P(r, z) + \phi_S(r, z) = 1 \tag{5.24}$$

where $\phi_{P,S}(r, z) = c_{P,S}(r, z)/\rho_{P,S}$ are the volumetric portions and $\rho_{P,S}$ are the densities of the polymer and solvent, respectively.

The simple calculations result in the following equation for variations in the local polymer concentration:

$$\Delta c_P(r,z) = \frac{\Delta\alpha(r,z)}{\varepsilon_P - \varepsilon_S(\rho_S / \rho_P)}$$

In order to estimate the initiation of a radial redistribution of the polymer concentration, the intensity profiles of the radiation transmission, $T_{exp}(r, z)$, were measured across the jet close to the jet beginning, as well as farther along the jet (see Figs. 5.6 and 5.7).

It turned out that close to the jet start, $T_{exp}(r, z)$ and $T_{sim}(r, z)$ almost coincide, as expected for homogeneous polymer solution (see Fig. 5.6), whereas farther along the jet, $T_{exp}(r, z)$ rises above $T_{sim}(r, z)$ at the jet center, indicating a change in the local absorption coefficient (see Fig. 5.7).

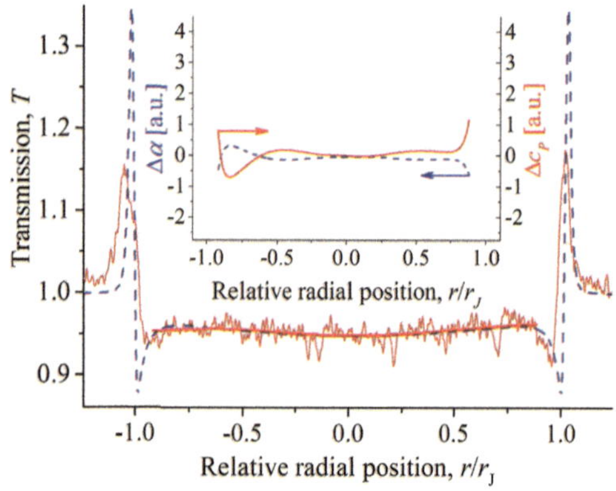

Figure 5.6 Typical absorption measurements across the electrospinning jet of a solution of 3 wt% PEO (M_w = 600 kDa) in water at the first *z*-position, indicated in Fig. 5.5b. The measured transmission T_{exp} (thin solid line) and its smoothing (thick solid line) compared to simulated transmission T_{sim} (dashed line). The inset displays calculated variations in X-ray absorption coefficient, $\Delta\alpha\ \phi_{P,S}(r, z)$ (dashed line), and resulting variations in the local polymer mass concentration, $\Delta c_P(r, z)$ (solid line). Z_1 = 0.02 mm, $r_J(z_1)$ = 120 μm. Reprinted from Ref. [74], Copyright 2011, with permission by the American Physical Society (https://doi.org/10.1103/PhysRevE.84.041806).

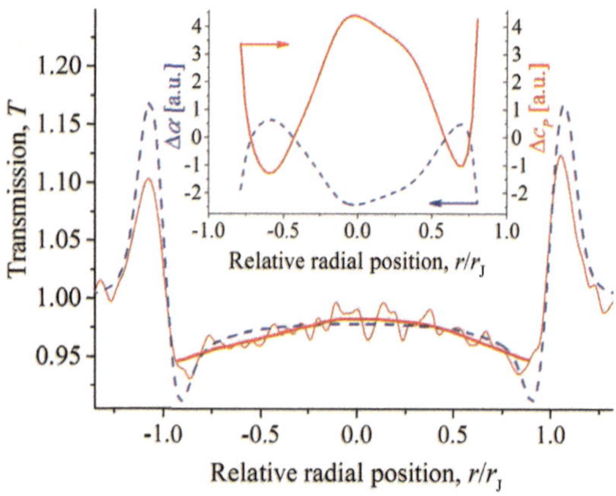

Figure 5.7 Typical absorption measurements across the electrospinning jet of a solution of 3 wt% PEO (M_w = 600 kDa) in water at the second *z*-position, indicated in Fig. 5.5b. The measured transmission T_{exp} (thin solid line) and its smoothing (thick solid line) compared to simulated transmission T_{sim} (dashed line). The inset displays calculated variations in X-ray absorption coefficient $\Delta\alpha(r, z)$ (dashed line), and resulting variations in the local polymer mass concentration, $\Delta c_P(r, z)$ (solid line). Z_2 = 0.5 mm, $r_J(z_2)$ = 22 μm. Reprinted from Ref. [74], Copyright 2011, with permission by the American Physical Society (https://doi.org/10.1103/PhysRevE.84.041806).

Note that in the aforementioned tests, the used polymer has a lower absorption coefficient than its respective solvent, so the decrease in absorption reflects a polymer concentration increase at the jet center.

5.2.2 Stretch-Induced Phase Separation in Polymer Solutions

Examining the instability in a jet of a viscoelastic semi-dilute entangled polymer solution under high stretching has also resulted in the same conclusion regarding a radial redistribution of the polymer concentration [75]. In doing so, this examining allows one to trace the system behavior up to the final evolution stage when polymer starts to concentrate in the jet center, forming a filament surrounded by the almost pure solvent.

It turned out that initially the analyzed jet was stable due to solution elasticity stopping the fluctuation growth (see Section 4.3 "Hydrodynamic instabilities of spinning jet", Eq. (4.13)). The point is that for the system analyzed in [75] (with concentration of about 20%, and $\gamma = 0.043$ N/m), the inequality (4.13) is true in the case of homogeneous transversal polymer distribution for jets with radii $r > r_{cr} \sim 1$ μm, whereas a decrease in the polymer concentration, for example up to 1% results in the increases of r_{cr} up to 100 μm, i.e., in this case, the capillary instability can occur in jets with the radius of about 100 μm. After some time due to a jet stretching, a radial gradient in the polymer distribution arose: The polymer started to concentrate in the jet center, whereas the solvent containing less than 0.5% of a residual polymer was remaining near the surface. In the case of such low polymer concentrations, the inequality (4.13) is violated, and that means the elasticity of such a solution cannot stop the fluctuation growth. The formed viscous liquid shell demonstrated Raleigh-type instability resulting in the formation of individual droplets on the oriented filament.

A snapshot demonstrating the solvent wringing out of a fiber and the formation of a near-surface layer of the solvent around an oriented polymer filament is depicted in Fig. 5.8a. The more prolonged evolution of the surface instability is shown in Fig. 5.8b. This process consists in the formation of an unduloid structure of low viscosity liquid on a cylindrical surface and finally its transformation of individual droplets.

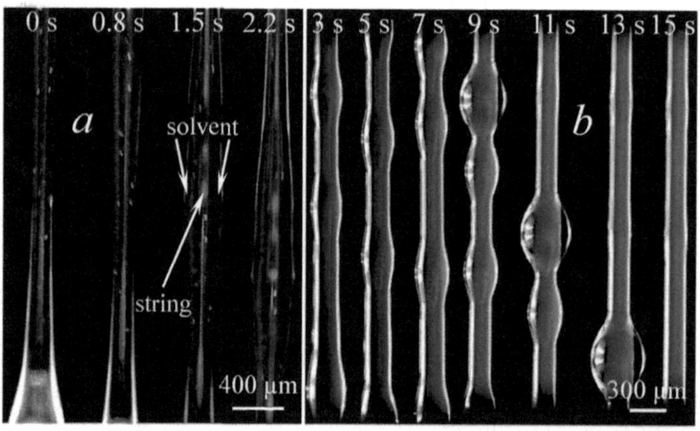

Figure 5.8 (a) Consecutive stages of "phase separation" accompanied by the formation of a near-surface solvent layer. (b) Appearance of separate droplets. The full observation time was 16 s. Noticeable waves appeared in the third second; droplets united with their neighbors and formed larger droplets in the seventh second; the large droplet descended along a string, collecting all solvent in the fifteenth second. Reprinted from Ref. [75] with permission of Springer. Copyright © 2014, EDP Sciences, SIF, Springer-Verlag Berlin Heidelberg.

The further system evolution results in the formation of the so-called "bead-on-a-string structure" consisting of droplets situated on the thin filament in the center of the jet (see Fig. 5.9). The feature of this structure is that the beads are solutions containing relaxed macromolecular coils, while macromolecules in thinning bridges are highly stretched and oriented. The process resulting in the formation of a "bead-on-a-string structure" is called "blistering" (see, for example, [80, 81]), and was assumed to be caused by phase separation due to strong stretching of polymer solutions (such a phenomenon when strong stretching of high concentrated polymer solution leads to phase separation with liberation of separate drops of a solvent on a fiber surface was observed a long time ago [82]). Nevertheless, more detailed analysis of the process in question demonstrates that the observed effect cannot be explained only in the framework of thermodynamic phase separation. The point is that redistribution of the polymer across the stretching jet occurs continuously, demonstrating no jump-like behavior.

The droplet formation kinetics resulting in such a conclusion is presented in Fig. 5.9. In this series of experiments, a light beam was directed along the fiber axis from the bottom of a syringe. The jet is working as a light guide, and the optical heterogeneity (local change in the refraction index) allows us to visualize the boundary between regions with high and low polymer concentration (polymeric core and solvent).

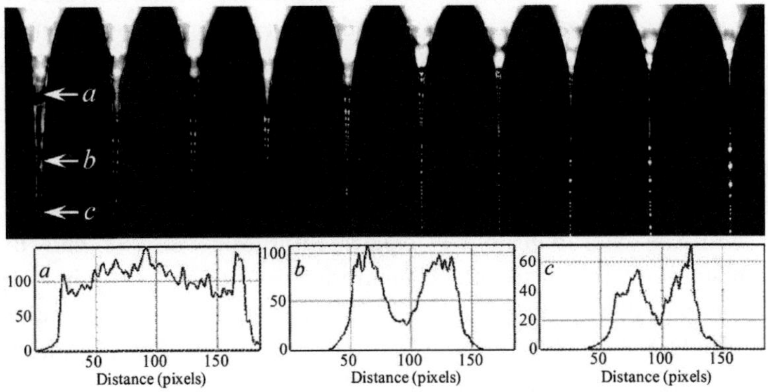

Figure 5.9 Consequent stages of the phase separation in stretching of an entangled polymer solution, accompanied by the formation of near-surface solvent layer and by appearance of separate drops. (PAN/DMSO – 20% solution). Time step equals 0.5 s. The light intensity showing the polymer distribution across the jet in three cross sections of the left snapshot is depicted in three bottom plots. Reprinted from Ref. [75] with permission of Springer. Copyright © 2014, EDP Sciences, SIF, Springer-Verlag Berlin Heidelberg.

Analyzing these images, we can conclude that a shrinkage of a polymer core occurs with stretching. The solvent release is being started in the form of a liquid cylinder at some distance from the Taylor cone vertex that corresponds to the definite extension ratio. The jet diameter is of about 90 μm at the distance equal to 2/3 of the full length and is being decreased along the jet up to 20 μm (see Fig. 5.9, the first snapshot), and this shrinkage is accompanied by polymeric core formation (see Fig. 5.9, plots a, b, and c). The next snapshots demonstrate the further reduction of the jet diameter. This decrease in the jet diameter is being accompanied by droplet initiation. Separate droplets are being determined quite well and can be seen by a naked eye. These droplets can be simply collected and taken off with a blotter paper. The chemical analysis of the collected droplets shows that, as mentioned earlier, they contain no more than 0.5% of a residual polymer in comparison with 18–22% in the initial solution.

In addition, the droplets formed close to the cone vertex start to move back toward the cone against gravity and are dissolved in the unstretched part of a jet. Such a behavior has the simple explanation: the droplet motion is caused by gradient in capillary forces. Indeed, the Laplace pressure is increasing with a decrease in the jet diameter, giving rise to a gradient in the pressure which, in turn, results in the aforementioned movement of the droplets into the direction of an increase in the jet diameter. At the same time, in the central part of the jet where the jet diameter does not vary, the droplets do not move and remain on a core surface until they evaporate without dissolving the oriented polymer filament.

All the aforementioned experimental observations can be explained on the basis of the theoretical model proposed in [74] and details discussed in Section 5.1 "Theoretical model of stretching of a polymer topological network." The key point of this model is that a large longitudinal stretching of a polymer network results in its lateral contraction, so a solvent is pressed out of this polymer network because of the decrease of its volume. As a result, a filament covered by solvent shell is being formed. Very briefly, the stages of the polymer network evolution can be understood with the help of Fig. 5.10. Initially, in spite of the fact that the aforementioned polymeric network is a random one, the system in question can be modeled as a regular cubic lattice with an elementary cell having the scale of the mesh size ξ_0, determined in Eq. (5.1) (see Fig. 5.10a). If the jet stretching is not too high, the hydrodynamic contraction of the polymer network is dominant (see Fig. 5.10b). In the case of very high and, finally, almost maximum possible stretching, the polymer network contraction due to its stretching is dominant (see Figs. 5.10c and d, respectively).

Now it is suitable to focus on the final stages of the process in question, which correspond to the cases of very high and almost maximum possible stretching.

Introducing the strain degree as $\lambda = \xi_{||}/\xi_0$, Eq. (5.14) describing the relative contraction, $\xi_{\perp}/\xi_0$, of elementary cell versus strain, can be rewritten in the following form:

$$\frac{\xi_\perp(\lambda)}{\xi_0} = \frac{3}{2+\left\{1-\left[(a/\xi_0)(\lambda-1)\right]^2\right\}^{-1/2}}$$

$$= \frac{3}{2+\left\{1-\left[\phi_0(\lambda-1)\right]^2\right\}^{-1/2}}, \tag{5.25}$$

where $\phi_0 \propto (\alpha/\xi_0)^{3-1/v} = \alpha/\xi_0$ ($v = 0.5$). Equation (5.25) correctly describes the system behavior also at low strain ($\lambda \to 1$) when the relative contraction is also small ($\xi_\perp/\xi_0 \to 1$).

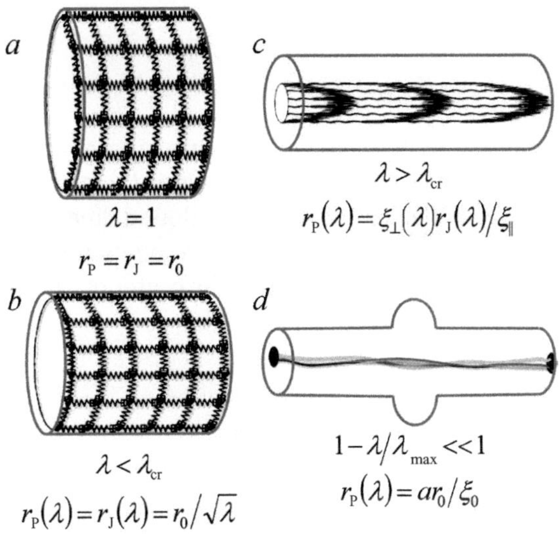

Figure 5.10 Schematic illustration of polymer network structure and its stretching. (a) Initial jet state; (b) low stretching of the jet; (c) high stretching of the jet; (d) formed filament and arising of capillary instability.

Now we have to take into account that in accordance with volume conservation law of incompressible liquid, the stretching of the jet is accompanied by its contraction:

$$r_J(\lambda) = \frac{r_0}{\sqrt{\lambda}}, \tag{5.26}$$

where r_0 and $r_J(\lambda)$ are the radii of unstretched and stretched jets, respectively.

This jet contraction results in the fact that the lateral contraction, $\xi_\perp$, of the mesh size is to be renormalized as follows:

$$\tilde{\xi}_\perp(\lambda) = \frac{3}{2 + \left\{1 - \left[\phi_0(\lambda - 1)\right]^2\right\}^{-1/2}} \frac{\xi_0}{\sqrt{\lambda}} \tag{5.27}$$

The effective polymer network radius is also to be renormalized in the same way:

$$\tilde{r}_P(\lambda) = \frac{3}{2 + \left\{1 - \left[\phi_0(\lambda - 1)\right]^2\right\}^{-1/2}} r_J(\lambda)$$

$$= \frac{3}{2 + \left\{1 - \left[\phi_0(\lambda - 1)\right]^2\right\}^{-1/2}} \frac{r_0}{\sqrt{\lambda}} \tag{5.28}$$

The effective radius of the zone containing almost the entire polymer is shown in Fig. 5.11. A zone outside of this radius (between solid and dashed, dashed-dotted, and dotted lines in Fig. 5.11) contains almost pure solvent. So we have a situation similar to a phase separation: oriented polymer in the center of a jet surrounded by a solvent (maybe by a dilute solution).

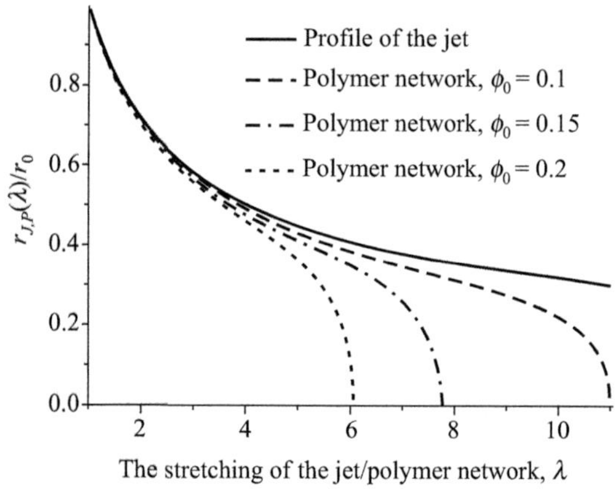

Figure 5.11 The lateral contraction of the polymer network, $r_P(\lambda)$, versus polymer network stretching, λ, with additional contraction due to hydrodynamic forces. The solid line corresponds to the jet profile $r_J(\lambda)$. The initial volumetric concentrations of polymer are: $\phi_0 = 0.1$, 0.15, 0.2. Reprinted from Ref. [75] with permission of Springer. Copyright © 2014, EDP Sciences, SIF, Springer-Verlag Berlin Heidelberg.

A nonlinear response of the polymer network under stretching results in changes in its volume that, in turn, results in changes in local polymer concentration. Indeed, the volumetric polymer concentration in a stretched system is

$$\phi(\lambda)=\frac{V_{\mathrm{P}}}{V_{\mathrm{Cell}}(\lambda)}=\frac{V_{\mathrm{P}}}{V_{\mathrm{Cell}}^{(0)}}\frac{V_{\mathrm{Cell}}^{(0)}}{V_{\mathrm{Cell}}(\lambda)}, \tag{5.29}$$

where the initial cell volume is $V_{\mathrm{Cell}}^{(0)}=\xi_0^3$, and the volume of the stretched cell is $V_{\mathrm{Cell}}(\lambda)=\tilde{\xi}_{\perp}^2(\lambda)\xi_{\parallel}$. The first factor in Eq. (5.29) corresponds to initial polymer concentration, ϕ_0; and the second one can be calculated with the help of Eq. (5.27):

$$\phi(\lambda)=\frac{\phi_0\xi_0^2}{\lambda\tilde{\xi}_{\perp}^2(\lambda)}=\frac{\phi_0}{9}\left\{2+\frac{1}{\sqrt{1-\left[\phi_0(\lambda-1)\right]^2}}\right\}^2 \tag{5.30}$$

Using the obtained dependence (5.30), the volumetric portion of residual solvent, $1-\phi(\lambda)$, remaining in the stretched network is depicted in Fig. 5.12. The vertical dashed lines indicate the maximum possible stretching of the system in question (for given initial polymer concentrations) with no disentanglement of the topological network. In the case of relatively low deformations, the solvent content is decreasing slowly, but beyond some threshold, corresponding to almost maximum possible stretching (about 90% of the maximum), the volume content of residual solvent sharply decreases and tends to zero. In other words, under a very high draw ratio, the solvent is " wringing out" of the polymer network, resulting in oriented filament formation and polymer solidification. So this model can be called the model of "wringing out a wet rag" .

Now using the experimental parameters of the system in question, one can estimate its maximum possible draw ratio. According to experiment conditions, the normal stresses, σ, at the syringe output created by the jet weight is of about 10^3 Pa. The rubbery shear modulus E_S of the solution is of about 20 ÷ 30 Pa. Then, according to the basic equation of rubbery elasticity, $\sigma=E_S(\lambda^2-1/\lambda)$, the draw ratio corresponding to the intensive phase separation is close to 6 ÷ 7, which agrees well to the theoretical estimation (see Fig. 5.12).

Finally, note that although the filament formation under high draw ratio seems similar to a phase separation process, in our opinion, polymer densification due to a lateral contraction of the polymer network being accompanied by the solvent replacement is dominant. Phase separation (if this term can be used) occurs only in the last stage of the process (see Fig. 5.10d).

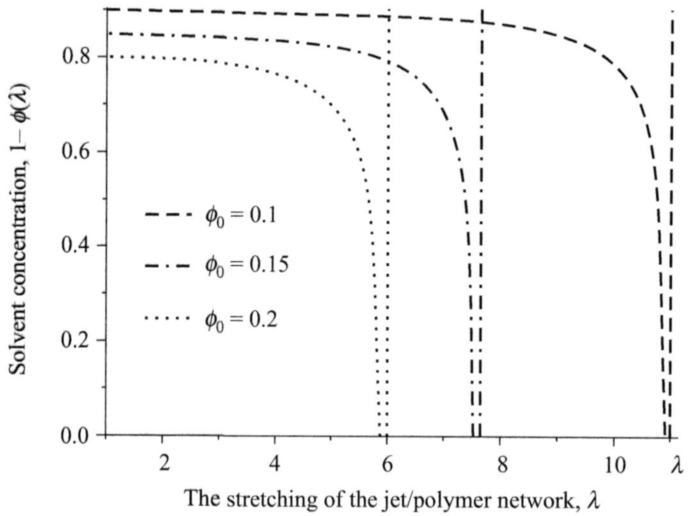

Figure 5.12 The volumetric portion of residual solvent, $1 - \phi(\lambda)$, remaining the forming filament versus polymer network stretching, λ (the vertical lines show the maximum possible stretching). The initial volumetric concentrations of polymer are: ϕ_0= 0.1, 0.15, 0.2. Reprinted from Ref. [75] with permission of Springer. Copyright © 2014, EDP Sciences, SIF, Springer-Verlag Berlin Heidelberg.

Summarizing results of the two last experimental studies, we can conclude that the fabrication of homogeneous nanofibers is possible when the spinning conditions provide enough strong longitudinal polymer network stretching being accompanied by radial contraction and formation of a filament in the center of a spinning jet. If the radial contraction of polymer network is faster than the thinning of the jet due to solvent evaporation, the formation of a solid skin fixing the fiber volume occurs at the final spinning stage when the amount of residual solvent in the system is not too high that, in turn, provides low level of free volume in the polymer matrix of electrospun nanofibers will be also enough. In this case, no voids or no porous structure will arise inside such fibers, i.e., the fabricated nanofibers will be homogeneous.

5.3 Disentanglement of Polymer Topological Network in Further Stages of Electrospinning

Further evolution of the polymer state inside a spinning jet is related to the reorganization of topological polymer network due to the macromolecule disentanglement. Such a point of view can be supported by the following arguments.

First, the simple estimation of shrinking of the radius of a spinning jet caused only by solvent evaporation and by aforementioned stretching of the topological network without its reorganization shows that the final diameter of as-spun fibers cannot be less than 10 μm. Indeed, the typical polymer concentrations, ϕ_P, of a spinnable polymer solution amount to about 10 ÷ 15 %. Therefore, as a result of only solvent evaporation, the fiber radius can be

$$r_{Ev} \propto r_{J0}\sqrt{\phi_P}\,,$$

where r_{J0} is the initial jet radius.

The aforementioned stretching of the topological network without its reorganization results in the elongation and shrinking of a polymer slug in $\sqrt{N_s}$ times, so the final fiber radius r_F can be

$$r_F \propto r_{J0}\sqrt{\phi_P / N_s} \approx 10 \div 30\ \mu\text{m},$$

where the initial jet radius r_{J0} is assumed to be about 1 mm; and the polymer subchains forming the topological network consist of 100 ÷ 1000 monomers.

The second argument in favor of the disentanglement process is related to the well-known rheological properties of polymers. A flow of the polymer melt (or concentrated solution) is possible due to disentanglement and re-entanglement of polymer macromolecules. In the case of high deformation rate, a liquid polymer system (melt or solution) transfers into the high-elastic state; moreover, when the deformation rate exceeds some critical value, a spurt (a stopping) in the flow of polymer solutions or melts occurs [83]. But even if the deformation rate of polymer is not too high, in the case of steady-shear flow, an entanglement loss (disentanglement) is also observed [84, 85]. The well-known polymer strain softening is also attributed to the macromolecule disentanglement [86].

The disentanglement process was considered a stage of polymer fiber formation by many authors. And even in one of the first papers devoted to electrospinning, Reneker and Chun have noted that "The ordered arrangements of molecules in the crystals eliminate entanglements and make it possible to convert the folded or extended chains in the crystals into smooth fibers by mechanical drawing" [8].

5.3.1 Multiple Necking

The further strain development results in necking or crazing and final rupture of the polymer matrix. Accepting this point of view, the next argument on the disentanglement process can be formulated: Under certain conditions (when the tensile stress is high enough), the macromolecule network disentanglement can result in the necking of the polymer nanofibers.

And, indeed, this phenomenon was observed experimentally. The multiple necking structure in the electrospun nanofibers collected on the rotating wheel was first reported by Zussman et al. [87] (see Fig. 5.13a). This phenomenon was attributed by authors to a strong stretching of solidified nanofibers by the accumulating wheel, if its rotation speed becomes too high. At the same time, the nanofibers collected on a grounded plate demonstrate no necking.

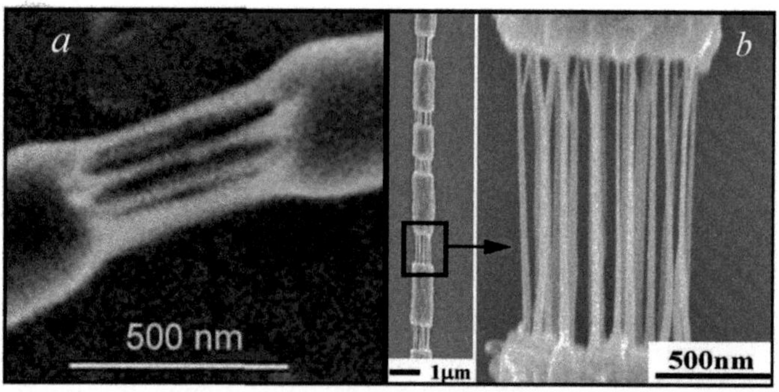

Figure 5.13 Typical images of necking in electrospun nanofiber. (a) Electron micrographs of fibrillar structures in 4% PEO electrospun nanofibers. The estimated diameter of the nanofiber is about 550 nm, whereas a single fibril diameter is about 20 nm. Reprinted with permission from Ref. [87], Copyright 2003, AIP Publishing LLC. (b) SEM images of a typical multiple necking fiber (left) and the magnified image (right) from the fibrillated part. Reprinted from Ref. [88], Copyright 2010, with permission from John Wiley and Sons.

Similar results were obtained by Yoshioka et al., attaining additional stretching of solidified fibers (in order to enhance the molecular orientation) by the use of a parallel-electrode collector [88] (see Fig. 5.13b).

5.3.2 Short Fiber Fabrication

The aforementioned observations indicate to one more possible result of the electrospinning process. Until now, we accepted the classification proposed by Shenoy et al. [53] according to which with the help of electrospinning, one can obtain (depending on the entanglement level of spinning polymer solution) good fibers, or beads (aerosol), or a mixture of fibers and beads (see Section 4.1 "Spinability of polymer solution," Fig. 4.6). But now we can conclude that the high strain rate of an extensional flow can result in fragmentation of a jet of a semi-dilute polymer solution, leading to the appearance of short nanofibers.

For the first time, short micro-fibers with the aspect ratio in the range of 10 to 200 were electrospun by Luo et al. [89], using a model polymer, polymethylsilsesquioxane (PMSQ) (see Fig. 5.14). It turned out that the aspect ratio and particularly fiber length are strongly influenced by the solvent system used for electrospinning and the molecular weight (M_w) of the polymer.

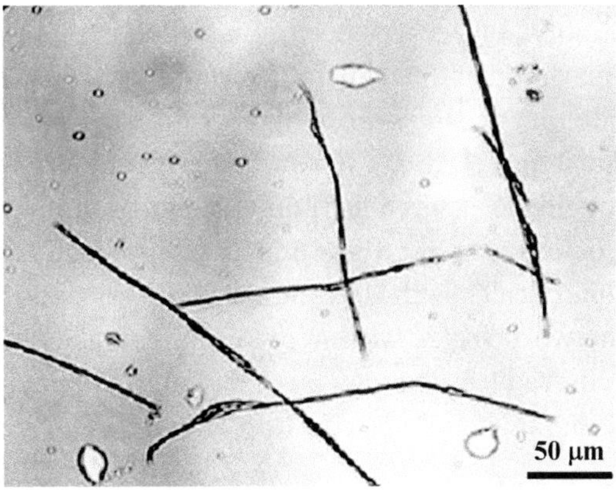

Figure 5.14 Optical micrographs (100× magnification) of electrospun short fibers from PMSQ solution (M_w = 7500) using methanol. Reprinted from Ref. [75] with permission of Springer. Copyright © 2011, Springer Science+Business Media B.V.

Thus, when using PMSQ with M_w = 7500 in methanol instead of acetone, short fibers with aspect ratio less than 200 were produced instead of continuous fibers (see Fig. 5.14). At the same time, when M_w of the polymer was decreased from 7500 to 4300, with all other conditions kept constant, a significant reduction in the aspect ratio of the as-spun fibers was observed. Short fibers with an average aspect ratio of 15 were produced from PMSQ solution with lower molecular weight (M_w = 4300) in 3:2 v/v dimethylsulphoxide:2-nitropropane. The average aspect ratio of short fibers spun from this PMSQ solution (M_w = 4300) in 2:3 v/v methanol:propanol was 31. However, PMSQ with high molecular weight (M_w = 7500) in both the aforementioned binary solvent systems produced long continuous fibers with aspect ratio > 3000 under the same spinning conditions.

Greenfeld and Zussman [90] have also demonstrated the possibility to fabricate short nanofibers with the help of the electrospinning process, focusing on the physical aspects of the process in question (see Fig. 5.15). In particular, they note that when the polymer network is not highly entangled, the high strain rate caused by the electrostatic field can stretch and disentangle chains from the network and break the jet into short segments. Entanglement loss is governed by two parameters: the entanglement number (the number of topological constraints along the chain) and the chain relaxation time; in doing so, both depend on the degree of polymerization, N, of the polymer chain. When the jet strain rate is low, rapid relaxation of the network prevents entanglement loss and viscosity is dominant. However, at high strain rates, relaxation is not sufficiently fast and elasticity is dominant, making chain extension and disentanglement possible. Taking into account the fact that the entanglement number, the number of topological constraints along the chain, scales with N, while the chain relaxation time in an entangled network scales with N^3, one can conclude that lower N (shorter chain) reduces entanglement, but at the same time reduces relaxation time; thus, the net effect on entanglement loss relies on the strain rate.

Thus, it is shown that short nanofibers can be created in a semi-dilute entangled solution, depending on the polymer architecture (linear or branched), molar mass, solution concentration and solvent quality. Also, short nanofibers are frequently accompanied by beads-on-a-string and free beads, a phenomena associated with low viscosity and high surface tension; and such a structure corresponds to the mixture of fibers and beads according to Shenoy classification.

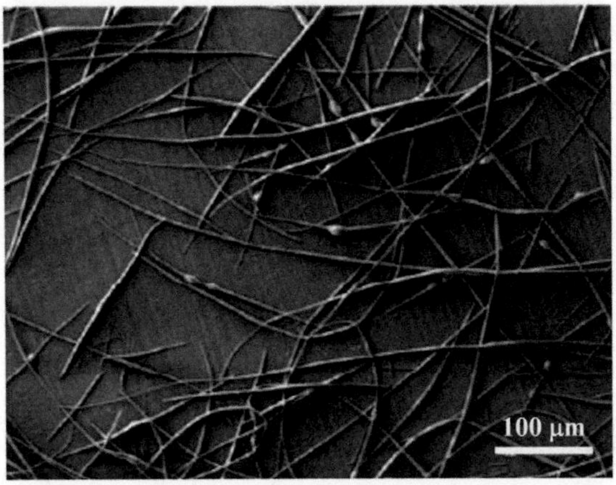

Figure 5.15 SEM micrographs of electrospun short nanofibers without beads-on-a-string structure. Electrospinning conditions were: molar mass 15 kDa PMMA, electric field 0.75 kV/cm, and flow rate 1 mL/h. Polymer concentration (vol %)/$CHCl_3$:DMF solvent composition (vol %) was 31.3/70:30. Reprinted from Ref. [90], Copyright 2013, with permission from John Wiley and Sons.

5.3.3 Simulation of Disentanglement Processes with the Help of a Pulley System

Now we would like to go back to the macromolecule disentanglement problem in order to discuss some physical features of this process being of high importance for our consideration.

Usually, the reorganization of polymer network occurs as a result of reptation (of slipping) of macromolecules due to thermal motion. However, the high rate and level of the system deformation, which are observed in electrospinning, result in the fact that in this case, the reptation of macromolecules is caused mainly by local stretching forces, whereas Brownian motion effect is negligible [91]. In doing so, the local pulling forces that provide the slipping of macromolecules are caused by hydrodynamic forces arising due to the velocity gradient along the spinning jet [74]. This fact allows one to utilize a pulley system for the description of polymer system evolution during electrospinning, more specifically, to describe the disentanglement processes in the polymer network.

Indeed, a probe macromolecule can be considered a rope passing through Pulleys. Each pulley, suspended by elastic spring, corresponds to an entanglement point (to a topological knot); and the springs on which the pulleys are suspended simulate the elasticity of the surrounding macromolecules (see Fig. 5.16a). Unifying all the pulleys into one, except the leftmost and the rightmost, the system can be reduced to the simplest one consisting of only three pulleys (see Fig. 5.16b).

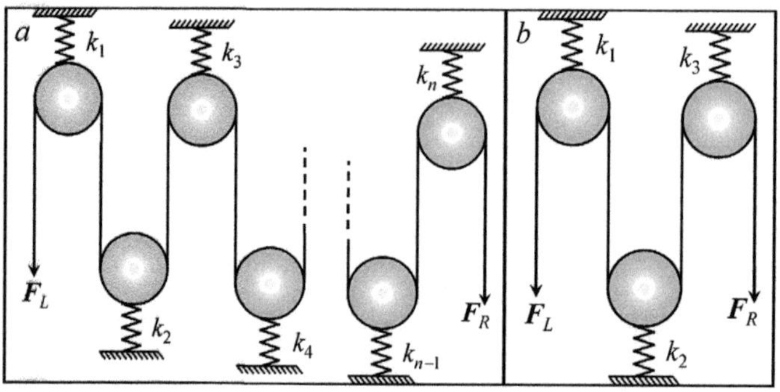

Figure 5.16 A pulley system utilized for examining the reorganization in the topological network of a semi-dilute entangled polymer solution: (a) the general case, (b) the simplest three-pulley system.

Thus, the minimal nontrivial system suitable for necessary analysis consists of three pulleys suspended by elastic springs, and of the inextensible rope of length L with two masses M_L and M_R on its ends, which simulate the hydrodynamic forces. Such a system can be fabricated very easily in any lab (and even at home) with the help of simplest elements: a rope, three rollers, three springs or elastic rubbery bands, and two masses (see Fig. 5.17a).

It turned out that the system behavior depends on the ratio of the used masses. If one of the masses is much larger than the other (for example, $M_R/M_L >> 1$), the rope can slip without any restrictions, and such a behavior corresponds to the full extrication of a macromolecule from an entanglement point, resulting in the disappearance of a topological knot (see Fig. 5.17b). If the difference in the masses is not too large, the pulleys start to approach each to other, while the slipping of the rope is considerably suppressed or fully stopped. In this case, we can say that the topological knots start to concentrate into bundles, forming strong-entangled regions (see Figs. 5.17c,d). Note that the intensive formation of the bundles of topological knots can be accompanied by system solidification (see Fig. 5.17d). And finally, one can vary the elasticity of the springs that were identical in all the above tests. It turned out that if the difference in the elasticity of the upper springs is large enough, the system behavior corresponds to the solidification of the polymer network (see Fig. 5.17e).

In addition, note that in the case of mobile pulleys, a rope motion is possible even if $M_L/M_R < k^3$ (where $k = \exp(-\mu\pi)$ and μ is the friction coefficient), whereas in the case of immobile pulleys, no motion is possible due to friction.

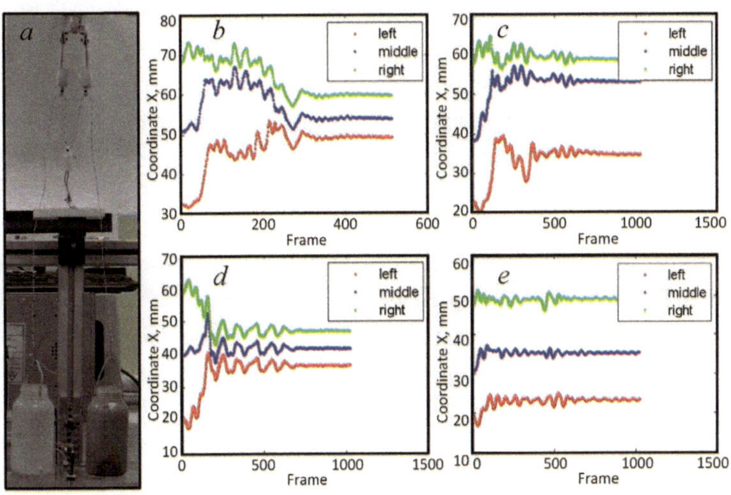

Figure 5.17 (a) An example of the three-pulley system fabricated in the lab, and the observed behavior of this system depending on the masses ratio; (b) a topological knot disappearance as a result of full extrication of a macromolecule from an entanglement point; (c) appearance of the bundle of the topological knots; (d) appearance of the bundle of the topological knots, accompanied by the system solidification, and (e) the system solidification in the case of various elasticity of the springs.

5.3.4 Spontaneous Symmetry Breaking in a Thermodynamic System of Pulleys

It seems that we have a simple macroscopic system that can simulate the disentanglement process and reorganization in the topological network of a semi-dilute entangled polymer solution, exposed to an essential stretching. However, a more detailed analysis of the system in question shows an additional problem. The point is that the necessary slipping and disentanglement of macromolecules are possible only if the difference in hydrodynamic forces applied to the ends of a macromolecule is larger than a critical value. Such a difference in hydrodynamic forces can appear only in the case of asymmetrical conformation of polymer macromolecules. But the amount of entanglement points per polymer chain in spinnable polymer solutions is large enough; therefore, the effective macromolecule conformations are symmetric. So we have to indicate a mechanism that can result in asymmetric macromolecule conformations. And only thereafter (arising of asymmetric macromolecule conformations) reorganization of the polymer network due to disentanglement processes can occur.

Thus, we have to analyze the behavior of the system consisting of three pulleys suspended by elastic springs and inextensible rope of length L with two masses M_L and M_R on its ends (see Fig. 5.18) [92]. In the case of symmetric system $M_L = M_R = M$ (see Fig. 5.18a). The friction between pulleys and rope should also be taken into account. Each pulley corresponds to a topological knot of a topological polymer network; the inextensible rope corresponds to a stretched reptated chain; the springs correspond to the surrounded polymer subchains; and the masses M_L and M_R correspond to the local forces occurring due to hydrodynamic interaction between the polymer network and solvent flow. The friction between pulleys and rope reflects the interaction of the probe macromolecule with the other ones in topological knots. If the symmetric system demonstrates a spontaneous symmetry breaking, we can turn to analysis of the pulley system with non-equal masses $M_L \neq M_R$ which describes the behavior of a macromolecule on the ends of which non-equal hydrodynamic forces are acting (see Fig. 5.18b).

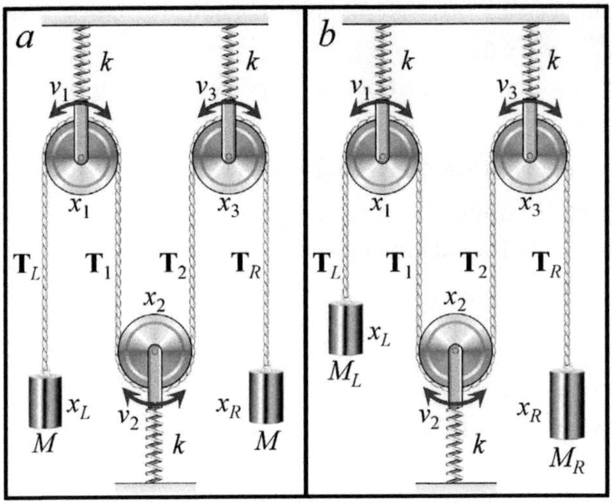

Figure 5.18 Schematic view of a pulley system in symmetric (a) and asymmetric (b) cases.

5.3.4.1 Mathematical description

The forces acting on the masses M_L and M_R are gravity and tensions in the rope, T_L and T_R, respectively. Therefore, the motion equations for the masses are

$$M_\text{L}\ddot{x}_\text{L} = T_\text{L} - M_\text{L}g, \tag{5.31}$$

$$M_\text{R}\ddot{x}_\text{R} = T_\text{R} - M_\text{R}g \tag{5.32}$$

The elastic forces of the springs are to be compensated by the tensions in the rope (the masses of pulleys and rope are assumed to be negligible compared to block masses):

$$K\Delta x_1 + T_\text{L} + T_1 = 0, \tag{5.33}$$

$$K\Delta x_2 + T_1 + T_2 = 0, \tag{5.34}$$

$$K\Delta x_3 + T_\text{R} + T_2 = 0, \tag{5.35}$$

where T_L, T_R, T_1, and T_2 are the tension forces in the rope in different legs of the system; $\Delta x_j = x_j - x_j^{(0)}$ are displacements of the pulleys (j = 1, 2, 3); k is the elastic coefficient of the springs; and g is the gravity constant (see Fig. 5.18). Note that the coordinate origin along the vertical axis is suitable to choose in such a way that $x_L^{(0)} = x_R^{(0)} = x_2^{(0)} = 0$.

The inextensible rope does not vary its length during any system evolutions:

$$(x_1 - x_L) + (x_1 - x_2) + (x_3 - x_2) + (x_3 - x_R) = L \tag{5.36}$$

If friction stops the rope slipping in the central pulley, the condition (5.36) should be replaced by:

$$2x_1 - x_2 - x_L = \delta + L/2, \tag{5.37}$$

and

$$2x_3 - x_2 - x_R = -\delta + L/2, \tag{5.38}$$

where δ is the initial displacement of the rope relative to the central pulley.

Using the following suitable re-notation $y_{1,3} = x_{1,3} \pm \delta - L/4$, the conditions for the inextensible rope can be simplified: the equations (5.36) can be rewritten as

$$2(y_1 + y_3 - x_2) - (x_L + x_R) = 0, \tag{5.39}$$

whereas Eqs. (5.37) and (5.38) can also be rewritten in the more simple form

$$2y_1 - x_2 - x_R = 0, \tag{5.40}$$

and

$$2y_3 - x_2 - x_L = 0 \tag{5.41}$$

The friction in the pulley results in the fact that the tension forces in the rope on the right and on the left from the pulley are unequal. In the case of the rope slipping, the ratio between these tensions (on the right and on the left from the pulley) can be calculated with the help of Eytelwein's equation. For example, in the case of the slipping in the left (or right) upper pulley, these ratios are

$$T_1 = \kappa^{sign(\dot{x}_1 - \dot{x}_L)} T_L \equiv \kappa^{\alpha_{1L}} T_L, \tag{5.42}$$

or

$$T_2 = \kappa^{sign(\dot{x}_3 - \dot{x}_R)} T_R \equiv \kappa^{\alpha_{2L}} T_R, \tag{5.43}$$

where $sign(x) = 1$, if $x > 0$, and $sign(x) = -1$, if $x < 0$; $k = \exp(-\mu\pi)$; μ is the friction coefficient; and $\alpha_{1L} \equiv sign(\dot{x}_1 - \dot{x}_L)$ and $\alpha_{2R} \equiv sign(\dot{x}_3 - \dot{x}_R)$

indicate the slipping direction. Note that if the following inequalities $\kappa T_{L,R} < T_{1,2} < \kappa^{-1} T_{L,R}$ are true, the rope will not slip through the corresponding pulley.

In the case of the slipping of the rope in the central pulley, the corresponding equation is

$$T_1 = \kappa^{sign\left[2(\dot{y}_3 - \dot{y}_1) + \dot{x}_L - \dot{x}_R\right]} T_2 \equiv \kappa^{\alpha_{12}} T_2, \tag{5.44}$$

where $\alpha_{12} = 1$ in the case of the rope motion from left to right, whereas in the opposite case, $\alpha_{12} = -1$.

Analogically, if $\kappa T_2 < T_1 < \kappa^{-1} T_2$, the rope will not slip through the central pulley. This inequality can be rewritten in the form $\kappa T_1 < T_2 < \kappa^{-1} T_1$, and such a transformation demonstrates the symmetry of the system in question.

5.3.4.2 Motion analysis

The variables $z_{p,m} = (x_L \pm x_R)/2$ describing displacements of the center of masses of the system and the difference in displacements of the left and right masses, respectively, are more suitable for our goal (to analyze the spontaneous symmetry breaking); so Eqs. (5.31) and (5.32) can be rewritten in the following form:

$$\ddot{z}_p = \frac{1}{M} T_p - g, \tag{5.45}$$

and

$$\ddot{z}_m = \frac{1}{M} T_m, \tag{5.46}$$

Where $T_{p,m} = (T_L \pm T_R)/2$, and z_m characterizes the system asymmetry.

It is clear that the tensions of the rope on the right and on the left from the pulley depend on the spring deformation; in doing so, the elastic force of the spring compensates the tensions in the rope. Using Eqs. (5.42) and (5.43), Eqs. (5.33) and (5.35) for the left and right upper pulleys can be rewritten as:

$$y_1 = -\frac{1}{k}(1 + \kappa^{\alpha_{1L}}) T_L, \tag{5.47}$$

and

$$y_3 = -\frac{1}{k}(1 + \kappa^{\alpha_{2R}}) T_R \tag{5.48}$$

In the case of symmetric evolution of the system with no slipping in the central pulley, the position of the central pulley, x_2, is calculated with the help of Eqs. (5.40) and (5.41), whereas in the case of asymmetric evolution of the system, when the rope slips in the central pulley, its position x_2 can be obtained with the help of Eqs. (5.34), (5.39), and (5.44):

$$x_2 = \frac{1}{k}(\kappa^{\alpha_{1L}} T_L + \kappa^{\alpha_{2R}} T_R) \tag{5.49}$$

Symmetric motion with no slipping in the central pulley.

First, we assume that in the case of no slipping in the central pulley, the masses are simultaneously moving in the identical directions (upward or downward), so the corresponding sign-functions are equal: $\alpha_{1L} = \alpha_{2R} \equiv \alpha$; in doing so, $\alpha = 1$ if the masses are moving downward, whereas in the opposite case when the masses are moving upward, $\alpha = -1$.

Substituting Eqs. (5.47) and (5.48) in Eqs. (5.40) and (5.41), we get the relationship between renormalized "tensions", $T_{p,m}$, and "displacements", $Z_{p,m}$:

$$T_p = -\frac{k}{2(1+2\kappa^{\alpha})} z_p, \tag{5.50}$$

and

$$T_m = -\frac{k}{2(1+2\kappa^{\alpha})} z_m \tag{5.51}$$

As a result, Eqs. (5.45) and (5.46) transform into oscillator equations:

$$\ddot{z}_p = -\omega_p^2(\alpha) z_p - g, \tag{5.52}$$

and

$$\ddot{z}_m = -\omega_m^2(\alpha) z_m \tag{5.53}$$

which have the following solutions, respectively:

$$z_p(t) = [z_p(0) + g/\omega_p^2]\cos(\omega_p t) + [\dot{z}_p(0)/\omega_p]\sin(\omega_p t) - g/\omega_p^2 \tag{5.54}$$

and

$$z_m(t) = z_m(0)\cos(\omega_m t) + [\dot{z}_m(0)/\omega_m]\sin(\omega_m t) \tag{5.55}$$

where $\omega_p^2(\alpha) = k/2(1+2\kappa^{\alpha})M$, and $\omega_m^2(\alpha) = k/2(1+\kappa^{\alpha})M$.

In the case of zero-initial conditions (symmetric state) $z_p(0) = z_m(0) = 0$ and $\dot{z}_p(0) = \dot{z}_m(0) = 0$, the solution (5.54) describing the oscillations of the center of masses of the system has the following simple form (see Fig. 5.19a, the first 5 seconds):

$$z_p(t) = -(g/\omega_p^2)[1 - \cos(\omega_p t)], \tag{5.56}$$

whereas Eq. (5.55) is being transformed into the zero solution: $z_m(t) \equiv 0$, i.e., no asymmetry appears in the system (see Fig. 5.19b, the first 5 seconds).

Note that due to the fact that $z_m(t) \equiv 0$, the tension forces T_L and T_R in the left and right legs of the rope are equal; therefore, the tension forces in the rope, T_1 and T_2, on the right and the left from the central pulley are also equal. That means our initial assumption that no slipping in the central pulley occurs is correct.

And now the assumption that $\alpha_{1L} = \alpha_{2R}$ is also to be checked. In the symmetric case, when $z_m(t) \equiv 0$, the functions describing displacements in the left and the right legs of the system are equal: $x_L(t) = x_L(t) = z_p(t)$, and $y_1(t) = y_3(t)$. Therefore, the arguments of the sign-functions α_{1L} and α_{2R}, as well as the functions α_{1L} and α_{1R}, themselves, are also equal.

Asymmetric motion with the slipping in the central pulley.

In the general case of asymmetric motion, the arguments of the sign-functions α_{1L} and α_{1R} are unequal:

$$\dot{y}_{1,3} - \dot{x}_{L,R} = -\frac{1}{2}\left(\frac{1+3\kappa^{\alpha}}{1+2\kappa^{\alpha}}\dot{z}_p \pm \dot{z}_m\right) \tag{5.57}$$

Nevertheless, the sign-functions α_{1L} and α_{1R} can be equal, if the following condition is satisfied:

$$\left|\dot{z}_m\right| < \frac{1+3\kappa^{\alpha}}{1+2\kappa^{\alpha}}\left|\dot{z}_p\right| \tag{5.58}$$

As our goal is to examine the symmetry breaking in the system, we can restrict ourselves by considering the case of low asymmetry, defined by the condition (5.58), with no analysis of more complicated system evolution when the motion directions of the rope in the left and the right upper pulleys are uncorrelated, or when the rope stops in one of these pulleys.

It turned out that the slipping in the central pulley is possible when the center of masses moves only in the upward direction when $\alpha = -1$ (for proof, see Eq. (5.65). In this case, the ratio between tensions T_L and T_R is:

$$T_L = \kappa^{\alpha_0} T_R, \tag{5.59}$$

where $\alpha_0 = \alpha_{12}$, i.e., $\alpha_0 = 1$, if the rope moves in the central pulley from left to right; and $\alpha_0 = -1$ in the opposite case.

The tension T_R can be determined with the help of the condition (5.39) as the following function of z_p:

$$T_R = -\frac{k}{(1+\kappa^{\alpha_0})(1+2\kappa^{\alpha})} z_p \tag{5.60}$$

As a result, the function z_p is described by the same equation (5.52), whereas Eq. (5.53) for the function z_m is to be modified:

$$\ddot{z}_m = \frac{1-\kappa^{\alpha_0}}{1+\kappa^{\alpha_0}} \omega_p^2(-1) z_p = -\frac{1-\kappa^{\alpha_0}}{1+\kappa^{\alpha_0}} (\ddot{z}_p + g) \tag{5.61}$$

This equation has nonzero solution (see Fig. 5.19b, the system behavior immediately after the 5th second):

$$z_m(t) = z_m(0) + \dot{z}_m(0)t - \frac{1-\kappa^{\alpha_0}}{1+\kappa^{\alpha_0}} [z_p(t) - z_p(0) - \dot{z}_p(0)t + gt^2/2], \tag{5.62}$$

where the time moment $t = 0$ corresponds to the start of the slipping.

Now we have to determine the sign of the parameter α_0, verifying thereby the self-consistency of the slipping condition. If the rope slips in the central pulley from left to right ($\alpha_0 = 1$), the length of the left part of the rope decreases; and in opposite case when the rope slips in the central pulley from right to left ($\alpha_0 = -1$), this length increases:

$$\frac{d}{dt}(2y_1 - x_L - x_2) \begin{cases} < 0, \text{if } \alpha_0 = 1, \\ > 0, \text{if } \alpha_0 = -1 \end{cases} \tag{5.63}$$

These two inequalities (5.63) can be reduced to the following:

$$\frac{1-\kappa}{1+\kappa} \frac{1+\kappa^{\alpha}}{1+2\kappa^{\alpha}} \dot{z}_p(t) + \alpha_0 \dot{z}_m(t) > 0, \tag{5.64}$$

where, like in Eq. (5.62), the initial time $t = 0$ corresponds to the time moment when the rope starts to slip through the central pulley.

If only symmetric evolution of the system precedes the motion with the slipping in the central pulley, the initial "asymmetric velocity" equals to zero: $\dot{z}_{\mathrm{m}}(0)=0$. In this case, in the time moment t = 0, which corresponds to the beginning of the slipping in the central pulley, the inequality (5.64) is simplified:

$$\frac{1-\kappa}{1+\kappa}\frac{1+\kappa^{\alpha}}{1+2\kappa^{\alpha}}\dot{z}_{p}(0)>0 \tag{5.65}$$

This simplified inequality (5.65) can be satisfied only in the case of positive $\dot{z}_{\mathrm{p}}(0)$, i.e., only if the center of masses moves upward, as was mentioned earlier. In the case of a downward motion of the center of mass ($\dot{z}_{\mathrm{p}}(0)<0$), the slipping condition (27) cannot be satisfied. Note that for upward motion, the parameter α indicating the movement direction in the left and right pulleys is negative: α = −1.

After changeover to asymmetric regime, the asymmetric motion continues for some time up to the time moment τ_{As}, at which the inequality (5.65) is broken:

$$g\tau_{\mathrm{As}}=\dot{z}_{\mathrm{p}}(0)-\frac{1}{2+\kappa}\dot{z}_{\mathrm{p}}(\tau_{\mathrm{As}}) \tag{5.66}$$

5.3.4.3 The energy balance

Thus, we have obtained that in the symmetrical case of equal masses (M_{L} = M_{R}), the initial motion can occur only without any slipping in the central pulley, and such a regime is observed only when the center of masses moves downward. But when the center of masses changes the direction of movement and starts to move upward, the slipping in the central pulley becomes possible. The changeover to a regime of motion with slipping in the central pulley is accompanied by a change in the elastic energy of the system in question (the kinetic and gravity energies do not change). Indeed, the changeover to the regime with slipping results in a change in the rope tension that, in turn, results in a change in the stretching of the springs in order to retain the forces' balance. As a result, the elastic energy of the system changes. In the case of symmetric motion, the elastic energy of the system is

$$U_{\mathrm{el}}^{\text{non-slip}}=\frac{k}{2(1+2\kappa^{\alpha})^{2}}\left[0.5(1+\kappa^{\alpha})^{2}+\kappa^{2\alpha}\right]z_{\mathrm{p}}^{2}, \tag{5.67}$$

whereas in the asymmetric case, this energy has the following form:

$$U_{\mathrm{el}}^{\mathrm{slip}} = \frac{k}{2(1+2\kappa^{\alpha})^2}\left[\frac{\left(1+\kappa^{2\alpha_0}\right)(1+\kappa^{\alpha})^2}{2(1+\kappa^{\alpha_0})^2}+\kappa^{2\alpha}\right]z_{\mathrm{p}}^2 \tag{5.68}$$

It turned out that as a result of the aforementioned transition, the energy can only increase. Indeed, subtracting Eq. (5.29) of Eq. (5.30), and taking into account that $\alpha = -1$, we find that after the changeover to a regime of motion with slipping in the central pulley, the change in the system energy is always positive:

$$\Delta U_{\mathrm{el}} = U_{\mathrm{el}}^{\mathrm{slip}} - U_{\mathrm{el}}^{\mathrm{non\text{-}slip}} = \frac{k}{4}\frac{(1-\kappa)^2}{(2+\kappa)^2}z_{\mathrm{p}}^2 > 0 \tag{5.69}$$

That means the changeover to a regime with slipping is accompanied by the overcoming of an energy barrier. The presence of this barrier stabilizes the mechanical system, and the slipping in the central pulley of the mechanical version of the system in question is impossible. Indeed, a changeover of mechanical system to a state with higher energy cannot occur without external action, which is accompanied by an increase in the system energy.

5.3.4.4 The thermodynamic system

The situation can dramatically change if the system in question is an element or a sub-system of some thermodynamic system. The point is that the elements or sub-systems of a thermodynamic system can have various energies, whereas in the case of a conservative system, the common energy, as well as its mean value corresponding to the system temperature T cannot be increased. (An increase in the system energy is possible only as a result of an external action that increases the system energy, like in the case of a mechanical system, but under such action, the system, getting the energy from outside, turns into a nonconservative one.) The probability to find an element or a sub-system of some thermodynamic system in the state with the energy E is proportional to $\exp(-E/k_{\mathrm{B}}T)$ (where k_{B} is the Boltzmann constant), i.e., decreases with the energy increase. In doing so, the energy of each individual element or sub-system can vary, whereas the common energy of the system cannot vary, remaining constant.

Taking into account the fact that the pulley system in question should simulate the behavior of individual polymer macromolecule, entangled in a network of other macromolecules, we can assume that the aforementioned pulley system is the sub-system of some thermodynamic system, and the common mechanical energy of the system can vary even with no energy swap with outside surrounding. This assumption results in the fact that in spite of the energetic barrier ΔU_{el}, (5.69), the changeover into a regime of motion with slipping in the central pulley can occur with a nonzero probability proportional to $\exp(-\Delta U_{el}/k_B T)$, i.e., spontaneous symmetry breaking is possible in a thermodynamic system, whereas in the mechanical system, the rope cannot start to slip in the central pulley under any conditions.

5.3.4.5 Numerical simulations

The evolution of the system in question was examined numerically. The system movement during the first half-period was described by the function $z_p(t) = -(g/\omega_p^2)[1-\cos(\omega_p t)]$, whereas $z_m(t) \equiv 0$.

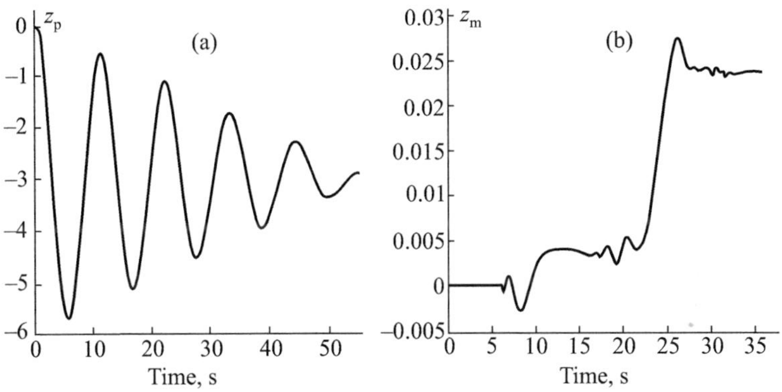

Figure 5.19 The system evolution obtained as a result of numerical simulations: (a) oscillations of the center of masses of the pulley system $z_p(t)$; (b) the nonzero dynamics of the function $z_m(t)$.

The first changeover to an asymmetric movement with slipping in the central pulley occurred in the beginning of the second half-period thanks to a nonzero changeover probability p at each calculation step, dt(p= 0.1 and dt = 0.001). This movement with slipping in the central pulley continued as long as the slipping condition was true. Thereafter, the regimes without and with slipping in the central pulley were alternating. In doing so, at the first changeover to a regime with slipping in the central pulley, the parameter α_0 was always equal to 1; at the next slipping regimes, this parameter was chosen randomly. The averaging over 100 described simulations for $\kappa = 0.9$ is presented in Figs. 5.19a,b; in doing so, the last one (Fig. 5.19b) demonstrates spontaneous symmetry breaking in the system in question.

Part III

Unique Properties of Electrospun Polymer Nanofibers

Chapter 6

Size-Dependent Behavior

6.1 Size-Dependent Behavior of Electrospun Nanofibers

6.1.1 Size-Dependent Behavior in Mechanical Properties

A great number of papers devoted to electrospun nanofibers often begin with a sentence similar to the following: "Polymer nanofibers, as well as polymer nanotubes, have attracted much attention recently because of their unique mechanical and thermodynamic properties compared to their bulk analogs." Such a statement is quite reasonable due to the fact that polymers in the form of electrospun nanofibers demonstrate noticeable improvement in their thermo-mechanical properties.

But similar outstanding mechanical and thermodynamic properties can be observed also in polymer fibers fabricated with the help of traditional technologies, with diameters achieving tens or hundreds of microns and even several millimeters. Indeed, such polymer fibers often demonstrate high tensile strength (more than 1 GPa), high ultimate elongation, good shape retention and crease resistance, and high resistance to repeated and alternating load. Such high physico-mechanical properties of polymer fibers can be achieved at formation and, especially, during post-processing by heat treatment, stretching, etc.

Electrospun Polymer Nanofibers
Arkadii Arinstein

ISBN 978-981-4745-27-7 (Hardcover), 978-1-315-36462-9 (eBook)
www.panstanford.com

Nevertheless, we have a serious reason to underline the uniqueness just of electrospun nanofibers as compared to the regular polymer fibers, fabricated with the help of traditional technologies. The point is that in addition to high physico-mechanical properties, electrospun nanofibers demonstrate a surprised feature: so called "size-dependent behavior" when the mechanical, thermo-mechanical, and thermodynamic properties of nanofibers depend on their diameter. Such a situation is impossible from the thermodynamics point of view, but multitude experimental studies demonstrate the size effect in the system in question. We discussed such a feature of electrospun nanofibers in Section 2.2 "Physical properties of nano-objects compared to the bulk," where some experimental observations of the size-dependent behavior of electrospun nanofibers in mechanical testings are presented: see Figs. 2.3 and 2.4 from Cuenot et al. [11, 14], Fig. 2.5 from Tan and Lim [15], Fig. 2.6 from Shin et al. [16], Fig. 2.7 from Ji et al. [17], and Fig. 2.8 from Arinstein et al. [12].

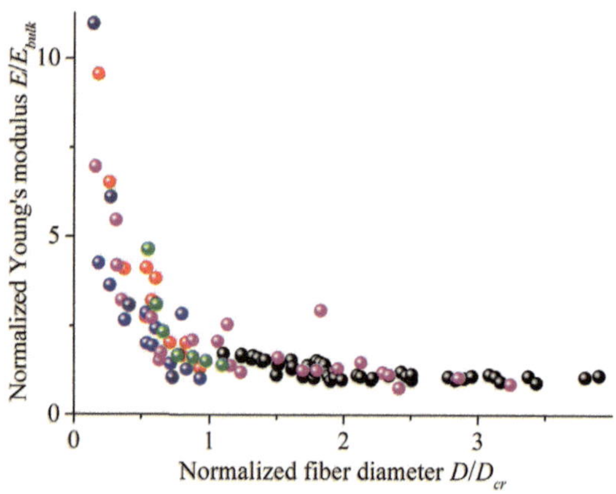

Figure 6.1 Elastic moduli, normalized by the bulk value, of different polymer nanofibers and polymer nanotubes as well as of metallic wires as a function of diameter, rescaled by their crossover diameters D_{cr}.

All the aforementioned famous examples of the size-dependent behavior demonstrate the same regular pattern: an abrupt increase in elastic moduli of different nano-objects when their transversal scale drops below a certain value, called a crossover scale. The fact that after a necessary renormalization, the dependences of elastic moduli of different nano-objects versus their diameters can collapse onto a universal curve is of high interest and requires a separate analysis and discussions. Such a universal curve is presented in Fig. 6.1 where the corresponding experimental data, published in several papers [10–12, 16, 17, 34], were rescaled by crossover diameter D_{cr} to collapse onto an universal curve.

Now let us draw our attention to one more remarkable fact: The molecular weight of polymer, which was used for nanofiber fabrication, also affects "size-dependent" elastic properties of electrospun nanofiber. It turned out that the relative moduli of polystyrene nanofibers of various molecular weights can be described with the help of one master curve as a function of fiber radius, scaled by R_g [10] (see Fig. 6.2).

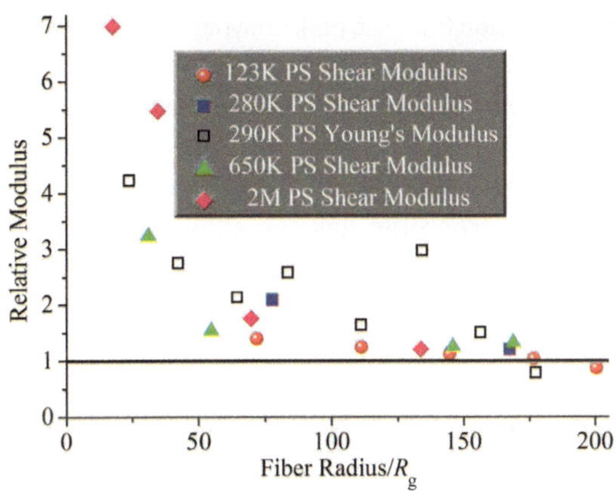

Figure 6.2 Relative moduli of fibers as a function of their radius, scaled by R_g. The distribution in diameter measurements was less than 10%. Reprinted with permission from Ref. [10], Copyright 2008, IOP Publishing.

This surprising observation is of high interest in light of the problem in question and will be discussed in Section 7.1 "Confinement mechanism of electrospun polymer nanofibers Reinforcement."

And finally, we would demonstrate here the set of experimental data obtained by Burman et al. [93] with the help of different experimental methods: tensile test [12], resonant vibration method [34], and three-point bending method [93] (see Fig. 6.3).

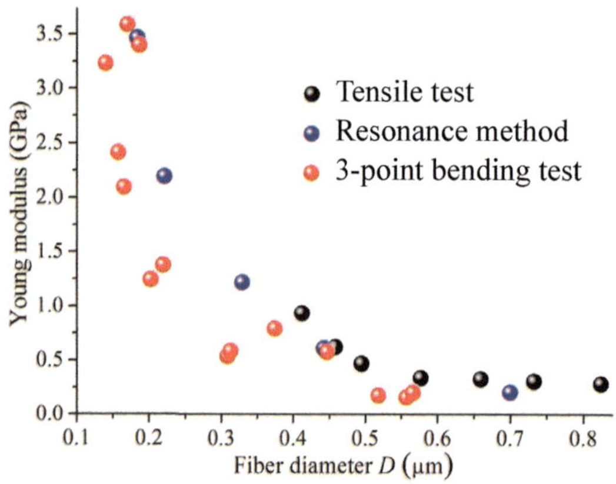

Figure 6.3 Dependence of elastic modulus E_{eff} versus nanofiber diameter D obtained with the help of different experimental methods: tensile test [12], resonant vibration method [34], and three-point bending method [93]. Reprinted with permission from Ref. [93], Copyright 2011, IOP Publishing.

One can see from Fig. 6.3 that all three testing methods resulted in the same values of measured moduli. Moreover, the DMA tests of the well-aligned mats also result in the same values of measured moduli, i.e., the macroscopic devices allow one to determine the thermo-mechanical parameters of electrospun nanofibers.

Thus, summarizing our speculations, we can conclude that all the aforementioned experimental data unambiguously indicate to the common feature of electrospun polymer nanofibers: Their elastic moduli start to depend on fiber diameters when these diameters become enough small, and with further decrease in fiber diameters, their moduli are increasing sharply. This feature of electrospun polymer nanofibers, called "size-dependent behavior," is nowadays the universally recognized fact. And in such a situation, we have to replace the question "Does size have an effect?" with "How can we explain these observations?"

6.1.2 Size-Dependent Behavior in Thermodynamic Properties

In discussing thermodynamic properties of electrospun nanofibers in the light of size-dependent behavior, first we have to remember a shift in their phase transition temperatures. As was mentioned in Section 2.2.2 "Thermodynamic properties," a depression of melting temperature as a function of the diameter of electrospun semi-crystalline PEO nanofibers was observed (see Fig. 2.11). The glass transition temperature T_g in electrospun PVA nanofibers can also decrease when their diameters decrease (see Fig. 2.12).

It turned out that in addition to the aforementioned shifts in melting and glass transition temperatures, variation of the surface properties of electrospun nanofibers can also be observed. For example, Stachewicz and Barber have shown that surface free energy of the electrospun polyamide 6 (PA6) nanofibers is almost 20% higher than the bulk PA6 cast film [94]. Such an increase in surface energy of the electrospun PA6 nanofibers was attributed by the authors of this paper [94] to the change in the ratio of polar and dispersive contributions into total surface energy of the electrospun PA6 nanofibers as compared to the bulk PA6 cast films (see Table 6.1).

Table 6.1 Surface free energy of solid surfaces, γ_s, and its dispersive, γ_s^d, and polar, γ_s^p, contributions

	γ_s, mJm^{-2}	γ_s^d, mJm^{-2}	γ_s^p, mJm^{-2}
PA6 film	40.34	34.9	5.5
PA6 nanofiber	47.0	5.2	41.8

Indeed, the necessary measurements and calculations allowed to demonstrate considerable increase in the polar component of the surface free energy of the PA6 nanofibers as compared with the cast film of the same polymer, whereas the dispersive component of the nanofiber surface energy had decreased. This phenomenon can be explained by an enhancement of the accessibility of polar oxygen groups at the electrospun nanofiber surface relative to the film that was demonstrated by chemical surface analysis using X-ray photoelectron spectroscopy. These results, in turn, confirm chemical group orientation at the electrospun polyamide nanofiber surface that promotes availability of polar groups for enhanced wetting behavior.

An increase in contact angle for different liquids on PA6 nanofibers as compared to PA6 cast films (see Table 6.2) also confirms the aforementioned conclusion.

Table 6.2 Contact angle for tested liquids on PA6 nanofibers, θ_n, and cast films, θ_f

Polyethylene glycol	43.4 ± 4.2	48.7 ± 2.9
Formamide	50.7 ± 3.5	62.8 ± 3.3

Thus, one can conclude that the electrospinning process affects the surface state of the polymer, increasing its surface free energy. "Whether the level of this increase depend on the nanofiber diameter?" is up to now an open question, but in our opinion, such an effect should be observed.

6.2 Do Surface Effects Explain the Unique Elasticity of Polymer Nanofibers?

Until recently, the unique mechanical and thermodynamic properties of nanofibers were attributed to the influence of surface effects [11, 95]. At first glance, such an explanation seems reasonable. Indeed, with a decrease in the specimen size, the surface-area-to-volume ratio is significantly increased and, in fact, is huge for nano-sized materials. Thus, it would seem obvious that surface phenomena have a dominant effect on nano-object properties. Moreover, Cuenot et al. [11] and He et al. [95] demonstrated that surface tension alone provides for the observed increase in the tensile modulus of Ag-nanowires. Nevertheless, a more detailed analysis of experimental data demonstrates that the influence of the surface effects cannot quantitatively describe the observed alterations in nano-object properties. It turns out that atomistic simulations of size-dependent elastic properties of nanowires [96, 97], as well as polymer nanofibers [98], show that surface influence is only observed in objects with diameters less than 5 nm. However, this scale (5 nm) is up to two orders of magnitude smaller than the samples for which experimental observations were obtained . It would seem that such quantitative inconsistency demonstrates a basic lack of understanding of the physical phenomenon in question, and an alternative interpretation of the aforementioned size-dependent behavior of nano-objects is necessary.

In order to quantitatively estimate the surface energy effect on the nanofibers's elastic modulus, the moduli of Nylon-6.6 polymer nanofibers obtained via three different experimental methods (three-point bending method [93], tensile [12] and resonant vibration [34] tests) were compared (see Fig. 6.3). The reason for such a comparison is that the surface energy affects the effective modulus only in the first method, which is based on bending deformation mode, and has no effect in the last two methods, which are based on tensile deformation mode.

Repeating the arguments presented in Section 3.2.2 "Tensile Test," we can conclude that the surface energy does not contribute to effective elastic modulus.

Indeed, the energy density estimation of uniaxial stretching of a fiber results in the quadratic function of the strain, $\Delta L/L$ (L is the initial fiber length), while the density of additional surface energy contains a positive linear and a negative quadratic term of the strain (the last one appears due to volumetric effects, and, for simplicity, it was not taken into account in Eq. (3.8))

$$U_{\text{tens}} = \frac{E}{2}\left(\frac{\Delta L}{L}\right)^2 + \frac{4\gamma}{D}(1-\nu)\left[\frac{\Delta L}{L} - \nu\left(\frac{\Delta L}{L}\right)^2\right], \tag{6.1}$$

where E is the material's elastic modulus, γ is the surface energy coefficient of a polymer, ν is Poisson's ratio, and D is the fiber diameter.

The stress (the first derivative of energy U_{tens}, (6.1)) caused by the elastic force is proportional to the fiber elongation. The elastic modulus E is the slope in the stress–strain dependence (the second derivative of energy U_{tens}, (6.1)) and contains a negative correction due to the surface effect resulting in some decrease in the measured modulus

$$E_{\text{eff}}^{\text{tens}} = E - \frac{8\gamma(1-\nu)\nu}{D} = E - E_{\text{sf}}^{\text{tens}} \tag{6.2}$$

Discussing the AFM three-point bending test, we found that the surface tension can affect the nanofiber modulus measured in the bending deformation mode (see Section 3.2.1 "Atomic Force Microscopy"). The point is that in the case of bending deformation, assuming that a force F applied at the clamped beam's midpoint induces a deflection, both the elastic energy and the additional surface energy of the bent beam are quadratic functions of the deflection δ

$$U_{\text{bend}} = -F\delta + \frac{1}{2}k\delta^2 + \frac{12\pi D}{5L}\gamma(1-v)\delta^2, \tag{6.3}$$

where $k = 192EI/L^3$ is the beam's elastic stiffness, $I = \pi D^4/64$ is the moment of inertia of the beam section, L and D are the fiber length and diameter, respectively, and v is Poisson's ratio [11].

Thus, the resulted elastic modulus is

$$E_{\text{eff}}^{\text{bend}} = E + \frac{8}{5}\gamma(1-v)\frac{L^2}{D^3} = E + E_{\text{sf}}^{\text{bend}}, \tag{6.4}$$

where $E_{\text{sf}}^{\text{bend}}$, which depends on γ, corresponds to the increased material stiffness arising from the creation of additional surface area during bending. Note, this analysis is valid for Euler–Bernoulli assumptions, where $\delta << D$.

The comparison of Eqs. (6.2) and (6.4) shows that the correction due to the surface effect in the case of pure tensile deformation is much smaller (in three orders of magnitude) than in the case of bending deformation: $E_{\text{sf}}^{\text{tens}} / E_{\text{sf}}^{\text{bend}} = 5v(D//L)^2 \leqslant 10^{-3}$. Therefore, in the case of pure tensile deformation, the surface effect is negligible and can be ignored ($E_{\text{eff}}^{\text{tens}} = E$), whereas in the case of bending deformation, the correction $E_{\text{sf}}^{\text{bend}}$ should be taken into account.

Thus, the data obtained from the three-point bending method, in which the measured effective modulus $E_{\text{eff}}^{\text{bend}} \equiv E_{\text{eff}}$ depends on the surface energy, and from the methods utilizing pure tensile deformation, in which the surface energy has negligible effect on the measured modulus, can be extrapolated, which clarify the role of the surface effect in the experimentally observed abrupt increase in the elastic modulus.

As can be seen in Fig. 6.3 where the test results of electrospun Nylon-6.6 polymer nanofibers obtained by three different methods are presented, an elastic modulus increase below a certain nanofiber diameter is observed. Similar results were also obtained by other researchers who performed tests utilizing only bending deformation mode [11, 95], and were attributed to the surface energy effect. However, the results that considered the data collected from the three-point bending method utilizing the bending deformation mode in which the measured effective modulus depends on the surface energy, as well as from the tensile [12] and resonant vibration [34] tests utilizing the bending deformation mode in which the measured effective modulus depends on the surface energy, are almost identical, but suggest an explanation contradictory to that stated by Cuenot et al. [11].

In order to clarify the surface energy effect, the experimental data obtained using the three-point bending method should be presented as follows:

$$E_{\text{eff}}D^3 = ED^3 + \frac{8}{5}\gamma(1-v)L^2 \tag{6.5}$$

Assuming D^3 as the argument x, this dependence can be interpreted as a linear function $y = \alpha x + b$, where the slope "α" corresponds to the material modulus, $\alpha = E$, and the shift along the y-axis corresponds to the surface energy, $b = \frac{8}{5}\gamma(1-v)L^2$. Note that Eq. (6.5) differs by the factor L^2 from the presentation of experimental data that was used by Cuenot et al. [11]:

$$E_{\text{eff}}\frac{D^3}{L^2} = E\frac{D^3}{L^2} + \frac{8}{5}\gamma(1-v) \tag{6.6}$$

As all fibers in the present three-point bending experiments were of the same length, minimizing the importance of L^2, Eq. (6.5) can be used to analyze data obtained from tensile and resonant vibration test scans (see Fig. 6.4).

One can see that the data depicted in Fig. 6.4 cannot be approximated by a linear dependence. However, linear approximations can be separately used for the two sub-regions representing thin versus thick fibers (see dash-dot lines in Fig. 6.4). The crossover between these two regions corresponds to the crossover diameter D_{cr} in the dependence of elastic modulus E_{eff} on fiber diameter. At this point, the elastic modulus begins to increase (see Fig. 6.3). Indeed, according to Fig. 6.3, the crossover between bulk and size-dependent behavior occurs at $D_{\text{cr},1} \propto 0.5$ μm, and according to Fig. 6.4, the boundary between two sub-regions allowing linear approximations is situated at $D^3_{\text{cr},2} \propto 0.1\ \mu\text{m}^3 \approx D^3_{\text{cr},1}$.

Note that the slope of the linear regression for the second sub-region (thick fibers) corresponds to the bulk elastic modulus, and nonzero value of b demonstrates the effect of surface energy. At the same time, the fitting of the first sub-region (thin fibers) results in much higher slope, and $b \approx 0$ demonstrating the negligible effect of surface energy in comparison to the increase in the elastic modulus E of the fiber's polymer itself. These results can lead one to conclude that the elastic modulus E of thin nanofiber depends on the fiber diameter.

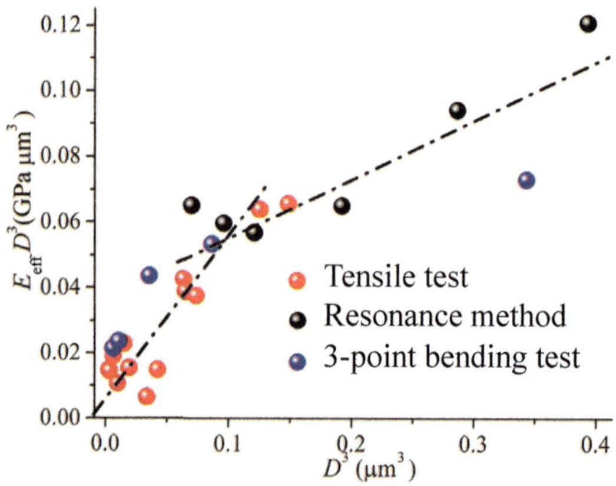

Figure 6.4 The dependence $E_{eff}D^3$ versus D^3 obtained using tensile test [12], resonant vibration method [34], and three-point bending method [93]. The data were fitted by linear regressions in two regions (dash-dot lines). Reprinted with permission from Ref. [93], Copyright 2011, IOP Publishing.

By using the last term in Eq. (6.4), one can calculate the direct contribution of surface energy by decoupling the measured modulus E_{eff} into the material modulus E and the additive modulus E_{sf}, which arises due to surface effects (see Fig. 6.5). The relative stiffness E_{sf}/E, as a function of the diameter of the tested nanofibers (for $\nu = 0.4$, and $\gamma = 47$ mN/m [99]), is presented in the inset of Fig. 6.5. It can be clearly seen that the relative contribution of additional stiffness of the polymer nanofiber material, due to the influence of surface energy E_{sf}/E_{mat}, decreases in direct correlation with changes in nanofiber diameter. These calculations demonstrate that for thin fibers, when an increase in the elastic modulus is most appreciable, the relative contribution of the surface energy to this change is only of ~10 %.

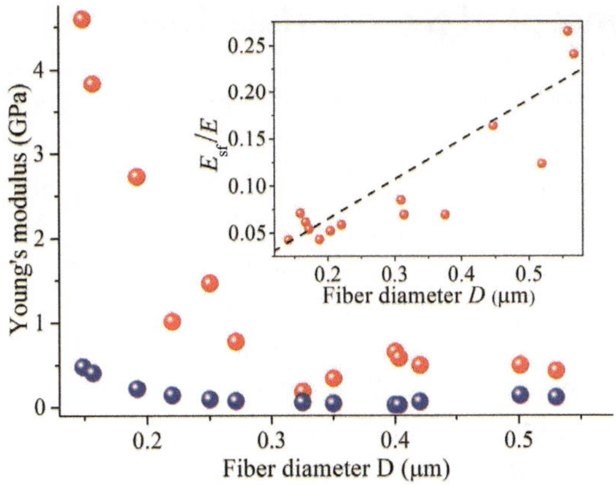

Figure 6.5 The elastic modulus E (red points) and the additive E_{sf} (blue points), which arise upon production of additional surface area during bending. The inset presents the ratio of the additive E_{sf} and of the material modulus E (E_{sf}/E_{mat}) versus nanofiber diameter D, obtained from a three-point bending test of polymer nanofibers [93].

Although surface energy is of great importance to the properties and behavior of nano-scale objects, the data presented here contradict the assumption that the observed increase in elastic modulus of polymer nanofibers is caused by surface energy effects. In other words, the aforementioned analysis of experimental data unambiguously shows that the abrupt increase in the elastic modulus of polymer nanofibers cannot be attributed to the influence of surface energy.

For completeness, it is appropriate to mention here that an upgrade of classical continuum mechanics (with the help of components containing derivatives of higher orders that take into account the local deformation gradients) was used to describe the size-dependent mechanical properties of nanofibers [100]. Without going into the details of such an approach, one can assert that only a physical mechanism can satisfactorily explain the size-dependent behavior of nano-objects, but the use of only a formal mathematical analysis of the problem in question will not provide an adequate explanation. Therefore, alternative explanations having no relation to surface energy need to be proposed.

6.3 Concept of Near-Surface Layer

Attempts to explain the size-dependent properties of polymer nano-objects by correlating them to the peculiar state of macromolecules in the near-surface layer [10, 101–104] cannot be evaluated due to the complexity of producing quantitative results in the size range in which these phenomena are observed. Indeed, the thickness of this near-surface layer does not exceed a few nanometers, while the diameter of nanofibers demonstrating a modulus increase can be up to 500 nm, exceeding the near-surface layer thickness by two orders of magnitude. A similar discrepancy is observed between the thicknesses of the near-surface layer and that of the metallic nanowires/films demonstrating a modulus increase [105, 106]. When expressing these considerations in terms of effective surface energy, the deficiency in quantitative estimations can be possibly explained.

Nevertheless, basing on their previous findings that a considerable polymer chain alignment in the near-surface layer of fiber occurs, Stachewicz et al. [107] proposed to explain an increase in the fiber elastic modulus (see Fig. 6.6) by the impact of the above aligned near-surface layer, assuming that elasticity of this aligned layer is much higher as compared to the bulk.

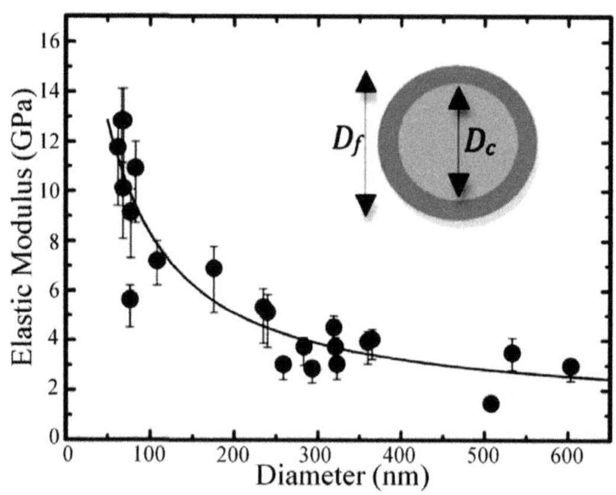

Figure 6.6 Plot of the variation in elastic modulus with electrospun PVA fiber diameter. The line shows a fit of Eq. (3) to the bending data for a fiber diameter D_f and a core diameter D_c. Reprinted from Ref. [107], Copyright 2011, with permission from Elsevier.

The arguments of the authors of the paper [107] are as follows:

1. As the elastic modulus of individual electrospun PVA nanofibers was measured using AFM-based bending tests, the elastic modulus of individual electrospun PVA fibers is being determined, using elastic beam bending theory (see previous Section 6.2 "Do surface effects explain the unique elasticity of polymer nanofibers?"):

$$E = \frac{F}{\delta}\frac{L^3}{192I} \tag{6.7}$$

where F is the applied normal force, δ is the deflection at the midpoint of the freely suspended fiber length L, $I = \pi D^4/64$ is the second moment of inertia defined for a cylindrical beam with diameter D. (Note that Eq. (6.7) does not take into account the influence of the fiber surface tension.)

2. Electrospinning process affects the surface state of the polymer, improving the alignment of polymer macromolecules in the fiber near-surface layer (see Section 6.1.2 "Size-dependent behavior in thermodynamic properties"). Therefore, the fiber can be regarded as a composite structure consisting of an anisotropic surface region (near-surface layer) and a more isotropic core; and elastic modulus of such a structure can be calculated with the help of a simple rule of mixtures.

3. If the shell has a fixed thickness ΔD, in the case of bending deformation mode, this rule of mixtures results in the following equation determining the effective nanofiber modulus:

$$E_{\text{eff}} I = E_{\text{cor}} I_{\text{cor}} + E_{\text{sh}} I_{\text{sh}} \tag{6.8}$$

where $I_{\text{cor}} = \pi D_{\text{cor}}^4/64$ and $I_{\text{sh}} = \pi\left(D_{\text{cor}}^4 - D_{\text{sh}}^4\right)/64$ are the moment of inertia of the cross-sectional areas of the fiber core and shell, respectively (I is defined in Eq. (6.7)). The core diameter is related to the fiber diameter D by $D_{\text{cor}} = D - 2\Delta D$. Substituting I_{cor} and I_{sh} into Eq. (6.8), we get the following equation for effective nanofiber modulus:

$$E_{\text{eff}} I = \left(1 - \frac{2\Delta D}{D}\right)^4 E_{\text{cor}} + \left[1 - \left(1 - \frac{2\Delta D}{D}\right)^4\right] E_{\text{sh}} \tag{6.9}$$

which, in the case of thin and strong near-surface layer ($2\Delta D \ll 1$, $E_{\text{cor}} \gg E_{\text{sh}}$), can be simplified:

$$E_{\mathrm{eff}} I \approx E_{\mathrm{cor}} + \frac{8\Delta D}{D} E_{\mathrm{sh}} \tag{6.10}$$

The fit of Eq. (6.10) to the bending data is shown in Fig. 6.6 and yields values of E_{cor} = 1.2 GPa, E_{sh} = 17.8 GPa, and ΔD = 13.1 nm.

At the first glance, such speculations seem reasonable, explaining the reinforcement of the nanofibers by the growth of their surface-area-to-volume ratio. Unfortunately, the proposed model contains an internal contradiction. The point is that in the case of the tensile deformation mode, the measurements should result in the same values of the system parameters (see the conclusions of previous Section 6.2 "Do surface effects explain the unique elasticity of polymer nanofibers?"). But, the use of the rule of mixtures in the case of the tensile deformation mode requires a modification of Eq. (6.8), and this modification, in turn, results in a change in the obtained value of the system parameters.

Thus, the physical explanation of the size-dependent behavior is still required.

6.4 Confinement Concept

The lack of a successful explanation of the size-dependent behavior (see, for example, Fig. 6.1 in Section 6.1 "Size-dependent behavior in mechanical properties") in the framework of mechanical (macroscopic) concepts convinced us to look for a microscopic physical mechanism that can provide the phenomenon in question. For this goal, a detailed microscopic examination of the systems in question is needed. Analyzing the size-dependent behavior of some system, we have to conclude that, from the physical point of view, such a phenomenon should be associated with some internal system scale, which becomes comparable with its size when this system will be enough small.

The first idea is that the reinforcing phenomenon is attributed to supramolecular structures inside the nano-objects. Indeed, when analyzing the size-dependent behavior of some system, we have to conclude that, from the physical point of view, such a phenomenon should be associated with some internal system scale that becomes comparable with its size when this system will be small enough.

At the same time, the observed phenomenon can be attributed to a gradual ordering of the nanofiber microstructure as a function of the diameter as compared to bulk polymer. Indeed, if as a result of processing, the structure parameters (degree of crystallinity, orientation of crystallites along the fiber axis, and orientation of macromolecules of the amorphous part of the fiber) are noticeably higher in the nanofibers with diameters less than a determined critical value, then the abrupt increase in the elastic modulus can be attributed to a more ordered structure of the nanofibers.

However, the experimental examining of electrospun Nylon-6.6 nanofibers does not demonstrate any abrupt changes in the microstructure of the nanofibers as a function of their diameter. According to X-ray analysis, only a mild, monotonous increase was observed in the crystallinity and orientation of the nanofibers as a function of their diameter (see Fig. 6.7).

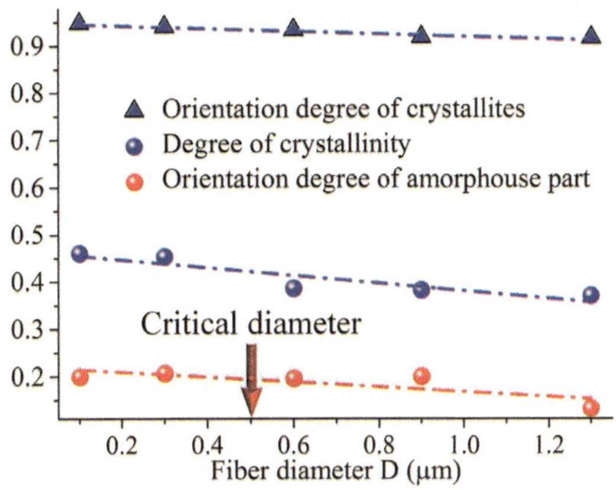

Figure 6.7 Orientation degree of crystallites along the fiber axis, degree of crystallinity, and order parameter of macromolecule orientation in the amorphous part of the fiber versus fiber diameter *D* of electrospun Nylon-6.6 nanofibers. Reprinted from Ref. [12], Copyright 2007, Nature Publishing Group.

It seems that this weak increase cannot be responsible for the dramatic change in Young's modulus of Nylon-6,6 nanofibers, observed experimentally (see Fig. 6.3 in Section 6.1, or 10 in Section 2.2.1 "Mechanical properties"). In addition, it turns out that the average size of crystallites does not depend on the nanofiber diameter and is only about 4 nm. In fact, the average size of crystallites in *all* nanofibers is much less than the fiber's diameter. In addition, as the crystallinity remains almost constant as a function of the nanofiber diameter, no significant change in the percolation structure (interconnected crystal structure that arises due to through-passed polymer chains) of the crystallites is expected.

6.4.1 Physical Model of the Phenomenon

Going one step further, we propose an alternative physical mechanism for the abrupt increase in the nanofiber elastic modulus. We began by examining a higher level of the nanofiber microstructure, namely, the supramolecular structure of the amorphous phase. Indeed, in elongation experiments, the stretching of polymer molecules (i.e., their orientational ordering) has a major effect on the increase in Young's modulus of the fibers, while the change in crystallinity of the polymer matrix has only modest effect. Therefore, we directed our attention to the supramolecular structures consisting of the oriented fragments of the polymer chains, and not to the crystals. Taking into account the fact that systems of semiflexible polymers and, especially, polymer fibers contain regions of oriented macromolecules (or of their fragments), we proceeded on the assumption that such supramolecular structures could explain the abrupt increase in the elastic modulus of nanofibers. Since no dramatic changes in nanofiber structure were observed, we directed our attention to dimensional effects. Therefore, the key question is whether the size of the oriented regions could be of the same order of the nanofiber diameter (see Fig. 6.8) and under what conditions would this be possible.

Many factors have an influence on the supramolecular structures of polymer fibers: the parameters of the polymer molecules, formation conditions, post-processing, etc. Thus, it would be difficult, if not impossible, to accurately estimate the size of a supramolecular structure. However, given the fact that the orientational ordering of chains in polymer fibers is not lower than in polymer melt, we can estimate the orientation correlation length in polymer melt to provide us with a lower bound for the required supramolecular structure size. Such estimation was already performed for the case of liquid oligomer systems in order to analyze the non-Gaussian conformational statistics of relatively short macromolecules [108, 109] using scaling concepts in polymer physics [110].

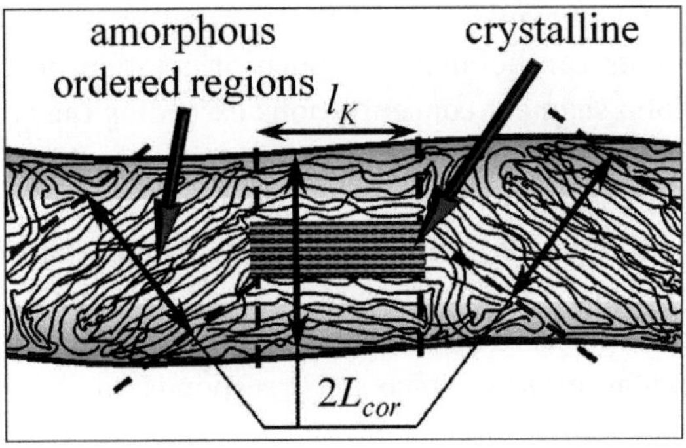

Figure 6.8 A sketch of the arrangement of crystallites and surrounded oriented amorphous matrix in polymer nanofiber, where L_{cor} is the orientation correlation length within the amorphous polymer portion, and L_K is the length of the Kuhn segment. Reprinted from Ref. [12], Copyright 2007, Nature Publishing Group.

6.4.2 Theoretical Estimations

It is well known that the volume occupied by a polymer chain (consisting of N links of size a) maybe estimated as $V_{ch} \propto a^3N^{3v}$ (v = 1/3 for a globule, v = 1/2 for a Gaussian ball, and v = 3/5 for a swelled ball). This volume, which is occupied by the trial chain, contains many polymer chains. According to the ergodicity hypothesis, which implies a self-averaging of a statistical system (in our case, ergodicity results in equiprobable spatial location of all polymer molecules), all these chains pass through the volume $V_K = V_{ch} / (N / n_K) \propto a^3 n_K^{3(1-v)+1} N^{3v-1}$ corresponding to a single Kuhn segment (a theoretical treatment allows one to present a real macromolecule as a polymer chain consisting of some freely joined straight segments, Kuhn segments, each of which contains n_K links). The volume V_K is filled with rods of length $l_K = an_K$ and diameter a, which are the Kuhn segments of other chains. The dimensionless density of these rods is $C \propto \phi/(a^3n_K)$, here ϕ is a specific volume of polymer with a value approximating one unit, $\phi \sim 1$.

Since the Kuhn segments of different chains are statistically independent, the above system is equivalent to the well-known system of rigid rods. Therefore, the model of Onsager [111] can be adopted for our analysis and the orientational ordering of the Kuhn segments can be analyzed. Such orientation ordering takes place for Kuhn segment concentrations exceeding the critical value $C \propto 1/al_K^2 = 1/a^3 n_K^2$ (this critical concentration value corresponds to that for a specific volume $\phi_{cr} \propto 1/n_K$). As we are dealing with polymers with relatively long Kuhn segments, the concentration of our system is much higher than the critical one, and orientational ordering takes place. The correlation length in such ordered systems is of particular interest since it corresponds to the size of the supramolecular structures in question.

The calculations performed in [108, 109] on the basis of the modified Onsager model [112], which corresponds to the case of high density of rods, produce the following value of the aforementioned correlation length:

$$L_{cor} \propto \frac{a}{\sqrt{\phi}}\left[n_K \ln(1-\phi)\right]^2 \tag{6.11}$$

Given the value of the correlation length and the value of the free volume in polymer systems, we can estimate the size of oriented regions in the polymer nanofibers. The free volume in polymer systems, $\vartheta = 1 - \theta$, can vary within a wide range. For example, in the amorphous phase, ϑ can amount to ~20%, but can decrease up to ~2% in crystallites [112]. Assuming a free volume of $\vartheta \approx 0.08$ (the arithmetic mean of the highest and lowest estimated values of the free volume provided by [113]), $a \simeq 0.1$ nm, and $n_K \approx 15$, the cross-sectional size of the ordered region is

$$D = 2L_{cor} \propto \frac{2a}{\sqrt{1-\vartheta}}\left[n_K \ln \vartheta\right]^2 = \tag{6.12}$$

$$\frac{2 \cdot 0.1 \cdot \left[15 \ln(0.08)\right]^2}{\sqrt{0.92}} \approx 300\,\text{nm}$$

This size amounts to a little more than half of the fiber diameter (see 2.8 in Section 2.2.1 "Mechanical properties") at which point, an abrupt increase in the fiber's elastic modulus is observed.

Nevertheless, this estimated lower bound can be assumed as a satisfactory value.

We can try to improve our estimation, taking into account additional factors influencing the formation of supramolecular structures in nanofibers. In particular, surface tension creates additional radial pressure in the fibers resulting in contraction of the polymer matrix; this produces a decrease in the free volume of the nanofiber and an increase in the specific volume. The additional pressure caused by surface tension forces in the case of a cylinder is $P_{ad} = \gamma/r$. The degree of restraint can be calculated as $\Delta r/r = P/E = \gamma/rE$ 0.0001 (where the following values of the relevant parameters were assumed to be $\gamma \propto 0.05$ N/m, $E \propto 2$ GPa, $r \propto 0.25$ μm, and for such parameters $P_{ad} = \gamma/r \simeq 0.2$ MPa or 2 atm). Taking into account that the volume reduction of the nanofiber during compression occurs primarily in the amorphous part of the polymer matrix, which constitutes about half of the fiber volume, we assume that $(\Delta r/r)_{amph} \approx 2\Delta r/r$. The compression of the polymer nanofiber results in a decrease in the free volume of the polymer matrix, but at the aforementioned system parameter values; the increase in the specific volume of the polymer molecules is negligibly small, $\hat{\theta} = \theta / \left[1 - (\Delta r / r)_{amph}\right]^2 \approx 0.924$, which results in only a 0.1% increase in the cross-cut size of an ordered region, $D \approx 300.5$ nm. This estimation demonstrates that surface effects have a negligible influence on the correlation length in the amorphous part of the polymer matrix.

The aforementioned arguments and considerable similarity between the estimation of the correlation length of ordered regions in nanofibers and the experimentally observed nanofiber critical diameter seem to validate the proposed concept, whose key point is that the scale of supramolecular structures in polymer nano-objects can be comparable to their characteristic size, and this fact results in noticeable variations in their mechanical and thermal properties.

Chapter 7

Explanations of the Size-Dependent Behavior by a Physicist: Some Possible Reasons and Mechanisms

7.1 Confinement Mechanism of Electrospun Polymer Nanofibers Reinforcement

As mentioned earlier, all proposed explanations of the size-dependent behavior of electrospun polymer nanofibers based only on the effects of surface tension [11] or near-surface layers [114, 115] cannot satisfactorily describe the aforementioned phenomenon quantitatively. Moreover, in contrast, the commonly held view that surface effects are dominant, an assumption that in polymer nanofibers the internal microstructure plays a more dominant role in the deformation process than previously thought [12], is quite reasonable. In accordance with this hypothesis, the dominant role in the polymer nanofiber reinforcement belongs to confinement-induced phenomena, and surface tension has a negligible effect. This assumption that the surface effect does not provide the size-dependent behavior of polymer nanofibers, in particular the abrupt increase in Young's modulus, was unambiguously proofed experimentally by Burman et al. [93] , who demonstrated that the impact of surface effect on polymer nanofiber reinforcement does

Electrospun Polymer Nanofibers
Arkadii Arinstein

ISBN 978-981-4745-27-7 (Hardcover), 978-1-315-36462-9 (eBook)
www.panstanford.com

not exceed 5 %, i.e., surface effect is negligible (see Section 6.4 "Confinement concept"). Note that McKenna et al. [116] have drawn the same conclusion in the case of the ultrathin cross-linked polymer films. The authors of this work argue that the surface tension contribution is insufficient to fully account for the observed overall rubbery stiffening behavior, which, actually, should be attributed to a mixture of both molecular stiffening and surface tension effects.

Coming back to the confinement hypothesis, note that from the physical point of view, the confinement-induced size-dependent behavior should be related to an internal scale parameter that is comparable with the scale of the system. Recently, Ji et al. showed that the relative elastic moduli of *polystyrene* nanofibers of various molecular weights can be described by one master curve as a function of fiber radius, scaled by R_g ($R_g \propto aN^{\nu}$ is the polymer radius of gyration, N is the number of monomers of a polymer chain, a is a monomer scale, $1/2 < \nu < 3/5$) [10] (see Fig. 6.2 in Section 6.1.1 "Size-dependent behavior in mechanical properties"). At first glance, this observation points to a polymer macromolecule as the structural element of electrospun nanofibers, which plays the dominant role in their size-dependent behavior, and the required scale corresponds to the macromolecule scale. However, according to these experimental data, the crossover to the size-dependent behavior occurs at $R/R_g \simeq$ 25–30 (where R is the fiber radius). Therefore, the radius of gyration, R_g, is too small to be assumed as the scale parameter comparable to the fiber radius. Nevertheless, the observed scaling phenomenon demonstrates that the aforementioned size-dependent behavior of electrospun polymer nanofibers is related to their internal structure.

Thus, further progress in the understanding of the features of electrospun polymer nanofibers is related to examination of their internal structure, which forms during fiber fabrication. The high strain rate (of the order 10^3 s^{-1}) acting on polymer solution during electrospinning [45, 58, 59] causes stretching and orientation of polymer chains, as indicated by in-process measurements of jets, using birefringence [68] and Raman [69] techniques. More detailed analysis of the dynamics of a highly entangled semi-dilute polymer solution under extreme longitudinal acceleration is presented in the work [74]. In particular, the theoretical modeling (see Section 5.1 "Theoretical model describing stretching of a polymer topological network"), supported by X-ray phase-contrast imaging of electrospinning jets of *poly(ethylene oxide)* and *poly(methyl methacrylate)* semi-dilute solutions (Section 5.2.1 “Radial contraction of an entangled polymer network during electrospinning”), has demonstrated that a substantial longitudinal stretching and transversal contraction of the polymer network caused by the jet hydrodynamic forces transform the network to an almost fully stretched state within less than 1 mm from the jet start. Although this almost fully stretched state will partially relax after jet solidification, when a stretching stress acting on the fiber sharply decreases, the final state of the polymer matrix of electrospun fiber is nonequilibrium, demonstrating some level of frozen supermolecular ordering.

There are two opportunities: either the fabrication of thinner fibers requires such modification of the spinning conditions, which results in an increase in the polymer matrix ordering; or the level of polymer matrix ordering does not depend on the thickness of electrospun nanofibers. The first opportunity was observed by Nakashima et al. for liquid crystal polymer nanofibers by WAXD measurements [117]. And the size-dependent behavior of these nanofibers was attributed by the authors to the ordering of the polymer matrix. Note that in this case, the type of polymer matrix ordering of electrospun nanofibers (formation of liquid crystals) corresponds to the equilibrium of the polymer in the bulk, and stretching during the spinning process additionally stimulates the ordering.

However, the opposite situation is also possible when the ordering of the polymer matrix of electrospun nanofibers does not depend on the fiber thickness. Our previous work showed that the electrospun Nylon-6.6 nanofibers demonstrate an abrupt increase in elastic modulus with a decrease in their diameters with no noticeable changes in their microstructure [12] (see Section sec7.5 “Confinement concept”). In this case, the amorphous portion of the polymer does not demonstrate any ordering in the bulk, whereas in the electrospun Nylon-6.6 nanofibers, the amorphous portion is partially oriented along the fiber axis. Nevertheless, this partial ordering of the polymer matrix that does not almost depend on fiber diameter cannot provide the phenomenon in question, and an additional physical mechanism resulting in the confinement-induced size-dependent behavior of polymer nanofibers is to be indicated.

Taking into account the fact that the size-dependent behavior was demonstrated by both amorphous and semi-crystalline polymer nanofibers, we can conclude that the amorphous portion of the polymer matrix of electrospun nanofibers plays a dominant role in the phenomenon in question. Therefore, we can restrict our consideration only by the theoretical model of size-dependent behavior of amorphous polymer proposed by Arinstein [118].

7.1.1 Structure of an Amorphous Nanofiber Polymer Matrix

Unfortunately, up to now, there is no detailed information regarding the structure of the polymer matrix inside electrospun nanofibers. We know only that the amorphous portion of the polymer inside electrospun fibers is partially oriented along the fiber axis.

Let us assume that the amorphous portion of the nanofiber polymer matrix contains anisotropic regions consisting of directional-correlated worm-like subchains, partially oriented along the fiber. These ordered regions have no clearly delineated boundaries and smoothly transfer one into another. Nevertheless, in order to introduce an effective internal structure of electrospun polymer nanofibers, these ordered regions can be assumed as ellipsoid-like "anisotropic particles" with one long, $l_{||}$, and two short, d_f, axes; in doing so, the long axis of these "anisotropic particles" is tilted relative to the fiber axis. The tilt angle θ is a random parameter being described by a distribution function. The mean value of the tilt angle Θ corresponds to the degree of orientational ordering in amorphous polymer matrix. The orientational ordering is characterized by the order parameter, $\eta = (3\langle\cos^2\theta\rangle-1)/2$, so $\cos\Theta = \sqrt{\langle\cos^2\theta\rangle} = \sqrt{(2\eta+1)/3}$. The typical values of the order parameter for polymer electrospun nanofibers are not too high. For example, the order parameter, η, [12] and $\Theta \approx \pi/4$, whereas for a completely disordered polymer matrix ($\eta = 0$), the mean value of the tilt angle is $\Theta \approx \pi/3$. We will use the mean value of the tilt angle Θ, and such approximation sufficiently simplifies mathematical calculations. Note that even if two neighboring "particles" have the same tilt angle, these "particles" can be nonparallel due to the possibility of free rotations in the plane perpendicular to the fiber axis and this fact is of high importance for our reasoning.

The aforementioned anisotropic ordered regions, separated by thin amorphous layers, play the role of structural elements reflecting the supermolecular structure of the polymer matrix of electrospun nanofibers. Note that the mechanical properties of the introduced "particles" are also anisotropic: In the direction of the long axis, the elastic modulus $E_{||}$ is much higher than in perpendicular directions $E_{\perp}$, and this fact is of great importance to our concept. It is reasonable to assume that $E_{\perp}$ is approximately equal to Young's modulus of amorphous polymer: $E_{\perp} \approx E_{am}$ ($E_{||} >> E_{am}$).

Thus, assuming the aforementioned anisotropic ordered regions as the structural elements of the polymeric matrix, one can depict the following effective internal structure of electrospun nanofibers: The nanofiber can be considered a "composite" consisting of effective "anisotropic particles" (the ellipsoid-like regions), surrounded by a binder (thin amorphous polymer layers) (see Figs. 7.1.a,b).

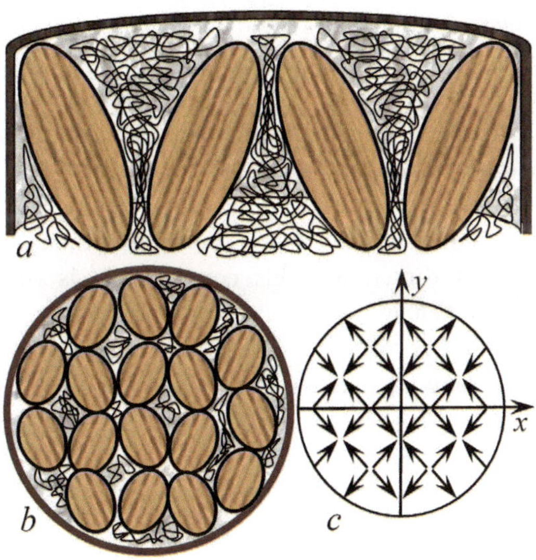

Figure 7.1 Schematic internal structure of as-spun polymer nanofiber. (a) A cross section along a fiber. (b) A cross section across a fiber. (c) The effective regular lattice of effective "anisotropic particles" across a fiber. Reprinted from Ref. [118], Copyright 2013, with permission from John Wiley and Sons.

The assumed detailed picture of the internal structure of electrospun polymer nanofibers is based on the following observations. Recently, Malkin et al. have shown that high rate deformations of the macromolecular network results in the formation of large inhomogeneous structures: "grains" or "bundles" [91]. Note that this phenomenon is in line with the well-known phenomenon called "shear-induced concentration fluctuations and/or phase separation," which has been examined both theoretically [119, 120] and experimentally [121]. Taking into account the fact that the electrospinning process is accompanied by high strain rate, it is reasonable to assume that similar structures can be formed also in electrospun polymer nanofibers. In addition, Tekoĝlut et al. have shown with the help of two-dimensional numerical simulations that granular solid demonstrates size-dependent behavior. In the case of shear and indentation deformations, the stiffness and strength of granular sample increase with decrease in its thickness [122].

Let us estimate the scale of the structural elements of the polymeric matrix of electrospun nanofibers.

According to the theoretical model describing the stretching and orientation of entangled polymer solution under inhomogeneous elongation [74] (see Section 5.1 "Theoretical model describing stretching of a polymer topological network"), the polymer subchains between two nearest (along the chain) topological knots stretch due to hydrodynamic flux with an increasing velocity; this stretching of subchains along the fiber axis results in their compression in perpendicular directions. This compression is related to the stretching of a subchain by the following equation (see Eq. (5.14) in Section 5.1.3 "Radial contraction of an entangled polymer network during electrospinning"):

$$\frac{\xi_{\perp}}{\xi_0} = \frac{3}{2 + \left[1 - \left(a\xi_{||}/\xi_0^2\right)^2\right]^{-1/2}} < 1, \tag{7.1}$$

where $\xi_0 \propto R_g\,(c/c^*)^{-1}$ is the mean mesh size between two nearest topological knots of polymer network in the equilibrium state, $R_g \propto aN^{\nu}$ is the radius of gyration of polymer coil, $n = (\xi_0/a)^2 = N(c/c^*)^{-2}$ is the monomer number in the aforementioned subchains, a is the scale of one monomer, c is solution polymer concentration, c^* is the crossover concentration of macromolecules overlap, N is the number of monomers of polymer chain (polymerization degree). Assuming $N = 10^3$, $c/c^* = 4$, we get $n \approx 50$; the ratios $\xi_{||}/\xi_0$ and $\xi_{\perp}/\xi_0$ can amount to $10 \div 15$ and 0.5, respectively, so the aspect ratio $\xi_{||}/\xi_{\perp}$ can amount to $20 \div 30$.

Such an aspect ratio allows one to consider the stretched subchains as worm-like objects and to apply the concept of orientational self-ordering to the system in question. The self-ordering of worm-like subchains, which is caused by the effects of excluded volume, in contrast to the hydrodynamic stretching, is not dominant, but this mechanism controls the ordering kinetics, determining the local scale of ordered regions within electrospun polymer nanofibers after solidification. The thickness of a bundle consisting of directional-correlated worm-like subchains inside a polymer solution jet can be estimated with the help of the following equation, obtained from Eq. 6.11 (see Section 6.4.2 "Theoretical estimations"):

$$d_s = \frac{\xi_{||}^2}{\xi_{\perp}\sqrt{\phi}}\left[\ln(1-\phi)\right]^2, \qquad (7.2)$$

where ϕ is the volumetric polymer concentration in the solution (the free volume in polymer solution, $\vartheta = 1 - \phi$, amounts about 0.9). After fiber solidification, the bundle thickness reduces up to:

$$d_f = \frac{a}{\sqrt{\phi}}\left[\frac{\xi_{||}}{\xi_{\perp}}\ln(1-\phi)\right]^2 \approx 3\,\text{nm} \qquad (7.3)$$

Note that the estimation of the correlation length, obtained in [12], exceeds the above bundle thickness d_f, by two orders of magnitude. This difference is caused by the fact that the bundle thickness (7.3) corresponds to a polymer solution correlation, frozen due to rapid solvent evaporation, whereas the correlation length, obtained in [12], corresponds to the condensed state of the semi-flexible polymer chain. Nevertheless, assuming the stretched subchain as a Kuhn segment, in the condensed state, the bundle thickness (7.3) has the same order of magnitude like the correlation length obtained in [12].

The length of the aforementioned bundle can be estimated assuming the random shift of neighboring worm-like subchains relative to each other in the fiber direction resulting in the diffusive law of the bundle length depending on the chain number in the bundle:

$$l_{\parallel} \propto \xi_{\parallel}\sqrt{\frac{d_{\mathrm{f}}}{a}} = \frac{\xi_{\parallel}^{2}\left|\ln(1-\phi)\right|}{\xi_{\perp}\phi^{1/4}} \approx 10\ \mathrm{nm} \tag{7.4}$$

Using Eq. (7.3) and (7.4), we find that the ordered regions are anisotropic with aspect ratio

$$\varepsilon = \frac{d_{\mathrm{f}}}{l_{\parallel}} = \frac{2\left|\ln(1-\phi)\right|}{(\xi_{\perp}/a)\phi^{1/4}} \approx \frac{1}{3} \tag{7.5}$$

Thus, according to our assumption, the nanofiber polymer matrix contains anisotropic regions consisting of directional-correlated worm-like subchains, partially oriented along the fiber; and the geometrical parameters of these ordered regions ("anisotropic particles") are described by Eqs. (7.3) and (7.4).

7.1.2 Mathematical Model for Polymer Nanofiber Elongation

At the first glance, the system of the introduced type can be analyzed on the basis of the theories of composite elasticity or of their further modifications [123, 124]. However, this analogy with a composite material is hardly applicable due to the extremely high concentration of the "anisotropic particles." Another opportunity related to granular materials or cellular solids seems more suitable. Indeed, the main distinction of granular materials from the system in question is weak linking between grains. Nevertheless, in the case of deformation mode at which this distinction is of no importance, the behavior of granular material and electrospun polymer nanofibers is similar. Unfortunately, we know no suitable models describing the aforementioned phenomenon in granular materials, so this analogy cannot help us to understand the physical mechanism of size-dependent behavior in nano-objects. Therefore, a mathematical model describing the evolution of the system in question under small deformation is required, and the necessary model will be formulated in this section as follows.

Let us examine a monolayer (in a cross section of the fiber) of the introduced "anisotropic particles." This monolayer having thickness

$l_0 = l_{||}\cos\Theta$ lies in the *xy*-plane (the *z*-axis is oriented along the fiber). In spite of the fact that these "anisotropic particles" are located randomly, in order to simplify the mathematical description, a regular square lattice can be introduced (see Fig. 7.1c). The "particles" are situated in the lattice nodes numbered by number pairs (I, j). In case of need, the irregular character of the system in question can be taken into account by introducing random fluctuations of system parameters.

Under external stretching force, the considered monolayer undergoes an elongation Δl, which is accompanied by relative rotations of the "anisotropic particles." The "particle" rotation occurs because of a torque arising due to local stress. This torque strives to decrease the angle between long "particle" axis $l_{||}$ and the fiber axis. At the same time, the relative displacement of neighboring "particles" gives rise to a torque having the opposite sign, so the rotation angle corresponds to the equality of these torques. Confinement effect is that the aforementioned rotations are hindered by the fiber surface layer where no rotations occur, and this fact is to be taken into account by boundary conditions. As a result, the elastic modulus depends on the diameter of the deformed fiber. In the case of small fiber diameters, this restriction is dominant while the effect decreases with increase in fiber diameter and tends to zero for large fiber diameters.

The elongation of the considered monolayer, Δl, is caused by both longitudinal elongation of the "anisotropic particles" and their rotations (an impact of a transversal deformation of the "particles" with low elastic constant E_{bulk}, into fiber elongation is negligible due to the fact that this deformation type is repressed by a counteraction of the neighboring particles). Thus, the fiber strain is

$$\varepsilon = \frac{\Delta l}{l_{||}\cos\Theta} = \frac{\sigma_{i,j}\cos\Theta}{E_{||}} + \delta\theta_{i,j}\tan\Theta, \tag{7.6}$$

where $\delta\theta_{i,j}$ is the rotation angle of the "particle" located in the point (x_i, y_j).

Equation (7.6) can be rewritten in the form of the relationship between the local stress $\sigma_{i,j}$ and the local rotation angle $\delta\theta_{i,j}$

$$\sigma_{i,j} = E_{\parallel}\varepsilon\frac{1-\left(\delta\theta_{i,j}/\varepsilon\right)\tan\Theta}{\cos\Theta}, \tag{7.7}$$

where the strain ε does not vary in fiber cross section.

The effective elastic modulus E_{eff} of the fiber is calculated as follows:

$$E_{\text{eff}} = \frac{\bar{\sigma}}{\varepsilon}, \tag{7.8}$$

where $\bar{\sigma}$ is the mean stress in the fiber cross section. Using Eq. (7.7), the mean stress $\bar{\sigma}$ can be calculated as

$$\bar{\sigma} = \frac{1}{S}{}_{i,j}\,\sigma_{i,j} = \frac{E_{\parallel}\varepsilon}{\cos\Theta}\left[1-\frac{\tan\Theta}{S}{}_{i,j}\,\frac{\delta\theta_{i,j}}{\varepsilon}\right] \tag{7.9}$$

Taking into account the radial symmetry of the system in question, the effective elastic modulus is

$$E_{\text{eff}} = \frac{E_{\parallel}}{\cos\Theta}\left[1-\frac{2\tan\Theta}{R^2}\int_0^R\frac{\delta\theta(r)}{\varepsilon}r\,dr\right] \tag{7.10}$$

Let us estimate the local torques acting on a trial "particle." The torque M_+ arises due to an external force action and, as was mentioned earlier, strives to decrease the angle between long ellipse axis $l_{\parallel}$ and fiber axis. It is proportional to the local stress $\sigma_{i,j}$; to the squared length of the long axis of the ellipsoid, $l_{\parallel}$; as well as to the first degree of the short one, d_{f}. In addition, this torque is equal to zero for the isotropic spherical particles, as well as in case when the angle Θ between long ellipsoid axis, $l_{\parallel}$, and the fiber axis is equal to 0 or $\pi/2$. So the torque M_+ can be approximated by the following equation:

$$M_+ \propto \sigma_{i,j}l_{\parallel}^2 d_f \sin\Theta\cos\Theta(1-\varepsilon) \tag{7.11}$$

There are also two torques caused by an elastic deformation and a shear stress, which arise as a result of a relative rotation of two neighboring ellipsoids. These two torques have the opposite direction relative to the torque M_+. The first torque M_{el} is proportional to the cubed length of the long axis of the ellipsoid $l_{\parallel}$ to $\cos^2\theta$ and to the rotation angle $\delta_{i,j}$. In doing so, the elastic constant is approximately equal to Young's modulus of amorphous polymer corresponding to bulk modulus E_{bulk}. So the torque M_{el} can be approximated by the following equation:

$$M_{\text{el}} \propto E_{\text{am}} l_{\|}^{3} \cos^{2}\Theta \cdot \delta\theta_{i,j} \tag{7.12}$$

The second torque M_{sh} is proportional to the length of the long axis of the ellipsoid $l_{||}$ to the cubed length of the short one, d_{f}, and to the differences in the rotation angles of the neighboring ellipsoids, $\delta_{i,j}$. In addition, this torque is inversely proportional to the thickness of a layer between neighboring particles, which consists of one or two polymer chains, so the thickness of this inter-layer can be taken as the scale of one monomer a. In this case, the elastic constant $l_{||}$ is related to the stretching of the tie molecules; therefore, $E_{||} >> E_{\text{bulk}}$. Thus, the torque M_{sh} can be approximated by the following equation:

$$M_{\text{sh}} \propto E_{\|} \frac{l_{\|} d_{\text{f}}^{3}}{a} \Big[\left(\delta\theta_{i,j} - \delta\theta_{i-1,j}\right) + \left(\delta\theta_{i,j} - \delta\theta_{i+1,j}\right) +$$

$$\left(\delta\theta_{i,j} - \delta\theta_{i,j-1}\right) + \left(\delta\theta_{i,j} - \delta\theta_{i,j+1}\right) \Big] \tag{7.13}$$

In the equilibrium, the torques acting in opposite directions have to compensate each other, and this condition allows one to write the following equation determining the internal state of the deformed fiber:

$$\sigma_{i,j} l_{\|}^{2} d_{\text{f}} \sin\Theta \cos\Theta (1-\varepsilon) = E_{\text{am}} l_{\|}^{3} \delta\theta_{i,j} \cos\Theta - \tag{7.14}$$

$$E_{\|} \frac{l_{\|} d_{\text{f}}^{3}}{a} \Big[\left(\delta\theta_{i+1,j} + \delta\theta_{i-1,j} - 2\delta\theta_{i,j}\right) + \left(\delta\theta_{i,j+1} + \delta\theta_{i,j-1} - 2\delta\theta_{i,j}\right) \Big]$$

This difference equation (7.14) can be approximated by the differential one:

$$\frac{1}{\xi} \frac{d}{d\xi} \xi \frac{d}{d\xi} \Phi(\xi) = \Phi(\xi) - \Phi_{0}, \tag{7.15}$$

where the dimensionless variable is used $\xi \equiv \xi_{I} = id_{\text{f}}/r_{0} = r/r_{0}$, $r_{0} = \left(d_{\text{f}}^{2} / \sqrt{a l_{\|}}\right) \left[(1-\varepsilon) \tan\Theta \sin\Theta + \left(E_{\text{am}} / E_{\|} \varepsilon \cos\Theta\right) \right]^{-1/2} \approx 10-20$ nm, and $\Phi_{0} = [\tan \Theta + (E_{\text{am}}/E_{||}\ \varepsilon(1 - \varepsilon) \tan \Theta)]^{-1}$. The boundary conditions for Eq. (7.15) are:

$$\left.\frac{d}{d\xi} \Phi(\xi)\right|_{\xi=0} = 0, \quad \Phi(R/r_{0}) = 0 \tag{7.16}$$

The solution to Eq. (7.15) satisfying the boundary conditions (7.16) is

$$\Phi(\xi) = \Phi_0\left[1 - \frac{I_0(\xi)}{I_0(R/r_0)}\right], \tag{7.17}$$

where $I_0(\xi) = J_0(i\xi)$ is the Bessel function.

Multiplying Eq. (7.15) by ξ and integrating it over ξ, we find

$$\frac{2}{R^2}\int_0^R \frac{\delta\theta(r)}{\varepsilon} r dr = \frac{2r_0^2}{R^2}\int_0^{R/r_0} \Phi(\xi)\xi d\xi$$

$$= \Phi_0\left[1 - \frac{2r_0}{R}\frac{d}{d\xi}\ln\left[I_0(\xi)\right]\Big|_{\xi=\frac{R}{r_0}}\right], \tag{7.18}$$

and the effective modulus E_{eff} is

$$E_{\text{eff}} = E_{\text{bulk}} + \Delta E \frac{1}{\xi}\frac{d\ln\left[I_0(\xi)\right]}{d\xi}\Bigg|_{\xi=\frac{R}{r_0}}, \tag{7.19}$$

where $E_{\text{bulk}} = E_{\text{am}}\cos\Theta/[\varepsilon(1-\varepsilon)\sin^2\Theta]$, $\Delta E = 2E_{||}/\cos\Theta$, and the following strong inequality is taken into account: $E_{\text{am}}\cos\Theta/[E_{||}\varepsilon(1-\varepsilon)\sin^2\Theta] << 1$.

The dependence (7.19), $E_{\text{eff}}(R)/E_{\text{bulk}}$, is depicted in Fig. 7.2 (dashed line).

Using the asymptotic behavior of Bessel function $I_0(\xi)$ for small and large arguments, one can find that for large fiber diameters ($R >> r_0$), the effective modulus E_{eff} increases with decrease in fiber diameter:

$$E_{\text{eff}} \approx E_{\text{bulk}} + \Delta E\left(\frac{r_0}{R}\right), \quad R >> r_0, \tag{7.20}$$

whereas for small fiber diameters ($R >> r_0$), the increase in the effective modulus E_{eff} stops, and it tends to a certain finite value

$$E_{\text{eff}} \approx \Delta E\left[1 - \frac{1}{8}\left(\frac{R}{r_0}\right)^2\right], \quad R << r_0 \tag{7.21}$$

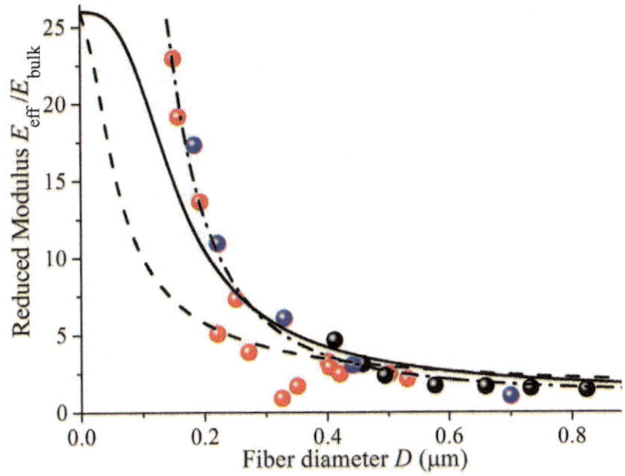

Figure 7.2 Dependence of relative Young's modulus E_{eff}/E_{bulk} versus nanofiber diameter D. The dashed line corresponds to Eq. (7.19) describing Young's modulus E_{eff} at r_0 = 10 nm. The solid line demonstrates the modified dependence $\hat{E}_{eff}$ (7.28) corresponding to the renormalization of the ratio R/r_0 (7.27) with $\tilde{a}N^{\gamma}$= 40. The ratio $\Delta E/E_{bulk}$ = 25. The dot-dashed line corresponds to the scaling approximation of the modified dependence $\hat{E}_{eff}$ (7.29) ($\tilde{a}N^{\gamma}$= 50 and $\Delta E/E_{bulk}$ = 50). The circles show the experimental values of relative Young's modulus E_{eff}/E_{bulk} for electrospun Nylon-6.6 nanofibers, obtained by the tensile test (black) [12], by the resonant vibration method (red) [34], and by the three-point bending method (blue) [93]. Reprinted from Ref. [118], Copyright 2013, with permission from John Wiley and Sons.

7.1.3 Scaling Properties of the Size-Dependent Elastic Modulus of Electrospun Polymer Nanofibers

Up to now, the normalization parameter r_0 has been assumed to be independent of the fiber diameter. But both the bundle thickness d_f and the bundle length $l_{||}$, which define the value of this parameter, depend on the volumetric polymer concentration in the solution c_V. In turn, the concentration of the spinning solution, c_V, affects the radius of electrospun nanofibers. Therefore, the normalization parameter r_0 depends on the radius of electrospun nanofibers, R, and this dependence will be defined below with the help of well-known scaling equations.

First, taking into account the fact that concentrations that are usually used in electrospinning are relatively low ($c_V \lesssim 0.15$), and assuming that ln $(1-x) \approx x$, Eqs. (7.3) and (7.4) can be rewritten as the following scaling dependences:

$$d_{\text{f}} \propto an^2\phi^{3/2},\ l_{||} \propto an^2\phi^{3/4}, \tag{7.22}$$

where $n = (\xi_0/a)^{\nu} = N(\phi/\phi^*)^{1/(1-3\nu)}$ is the monomer number in the subchains of the topological polymer network, $\phi^* = N^{1-3\nu}$ is the volumetric crossover concentration of macromolecules overlap. Therefore, the normalization parameter r_0 depends on the volumetric polymer concentration ϕ, as well as on the polymerization degree N, and this dependence also can be approximated by a scaling equation

$$r_0 \propto \frac{d_{\text{f}}^2}{\sqrt{al_{||}}} \propto a\left(n\phi^{7/8}\right)^3 = aN^{\alpha}\left(\frac{\phi}{\phi^*}\right)^{-\alpha/(3\nu-1)}, \tag{7.23}$$

where $\alpha = 3[1 - 7(3\nu - 1)/8]$.

It is well known that the radius of electrospun nanofibers depends on the concentration of spinning solution. Thus, Gupta et al. [52] showed that the relative concentration of spinning solution, $c_{\text{V}} / c_{\text{V}}^*$, and the final radius of electrospun nanofibers, R, are also related by a scaling equation

$$\frac{R}{R_0} \propto \left(\frac{\phi}{\phi^*}\right)^{\beta}, \tag{7.24}$$

where R_0 is the normalization parameter $R_0 \simeq 10^{-3} - 10^{-2}$ μm. The index β equals to about 3. According to the data, presented by Gupta et al., $\beta \approx 3.1$ [52]; the other measurements using a different type of polymer have shown the following values: $\beta \approx 3.0$ [56] or $\beta \approx 2.9$ [23] (see Section 7.2 "Shifting of the melting point for semi-crystalline polymer nanofibers", Fig. 7.6).

As a result, the normalization parameter r_0 can be presented in the following form:

$$r_0 \propto aN^{\alpha}\left(\frac{R}{R_0}\right)^{-\alpha/\beta(3\nu-1)} \tag{7.25}$$

Let us present the normalization parameter r_0 in the following form:

$$r_0 \propto aN^{\gamma}\left(\frac{R}{R_0N^{\gamma}}\right)^{-\alpha/\beta(3\nu-1)} \tag{7.26}$$

where $\gamma = \alpha\beta(3\nu - 1)/[\alpha + \beta(3\nu - 1)]$.

Such a view of the normalization parameter r_0 allows to present the ratio R/r_0 in the form of a power function of the fiber radius R:

$$\frac{R}{r_0} = \left(\frac{R}{\tilde{a}N^{\gamma}}\right)^{\delta} \tag{7.27}$$

where $\tilde{a} = (R_0/a)^{\gamma/\beta(3\nu-1)}\,a$ and $\delta = 1 + \alpha/\beta(3\nu - 1)$.

Substituting the ratio (7.27) into Eq. (7.19), we receive the modified equation for the effective module $\hat{E}_{\text{eff}}$:

$$\hat{E}_{\text{eff}} = E_{\text{bulk}} + \Delta E \frac{1}{\xi}\frac{d\ln\left[I_0(\xi)\right]}{d\xi}\Bigg|_{\xi=\left(\frac{R}{\tilde{a}N^{\gamma}}\right)^{\delta}} \tag{7.28}$$

In the case of large fiber diameters ($R >> \tilde{a}N^{\gamma}$), the effective modulus $\hat{E}_{\text{eff}}$ increases with fiber diameter decrease according to the power law

$$\hat{E}_{\text{eff}} \approx E_{\text{bulk}} + \Delta E\left(\frac{\tilde{a}N^{\gamma}}{R}\right)^{\delta}, \tag{7.29}$$

whereas for small fiber diameters ($R >> \tilde{a}N^{\gamma}$), the effective modulus $\hat{E}_{\text{eff}}$ tends to a certain finite value, also according to the power law with the doubled index

$$\hat{E}_{\text{eff}} \approx \Delta E\left[1 - \frac{1}{8}\left(\frac{R}{\tilde{a}N^{\gamma}}\right)^{2\delta}\right] \tag{7.30}$$

The values of the indexes γ and δ in Eqs. (7.27–7.30) cannot be considered free model parameters and are to be calculated. As the solvents usually used in electrospinning are good ones, the index ν can be estimated as $\nu \approx 0.55$. Assuming $\beta \approx 2.9$, we find that $\gamma \approx 0.75$ and $\delta \approx 1.6$.

The dependence (7.28) $\hat{E}_{\text{eff}}/E_{\text{bulk}}$, with $\tilde{a}N^{\gamma} = 40$, is depicted in Fig. 7.2 (solid line). In order to compare the dependence (7.19) and its modification (7.28), the values of the parameter $\Delta E/E_{\text{bulk}}$ were chosen the same in both cases: $\Delta E/E_{\text{bulk}} = 25$. One can see that the modified dependence (7.28) demonstrates better agreement with the experimental data compared with dependence (7.19), particularly for thin fibers, in region of small fiber diameters where the noticeable discrepancy between Eq. (7.19) and experimental data is being observed. Note that despite the fact that the dependence

(7.28), $\hat{E}_{\text{eff}}$, demonstrates physically reasonable behavior, remaining finite even if the fiber radius tends to zero, the proposed model is hardly applicable for such small radii. From this point of view, it is reasonable to use only the dependence (7.29), which agrees very well to experimental data (see Fig. 2, dot-dashed line).

The obtained result also allows to explain the experimental observation of Ji et al. [10] that the molecular weight affects the elastic modulus of electrospun nanofibers; and the relative elastic moduli of electrospun nanofibers of various molecular weights can be described by one master curve as a function of fiber radius, scaled by R_g, i.e., according to power law. Indeed, the argument R in Eq. (7.28) is scaled by the factor $\tilde{a}N^{\gamma}$ having the same form as the radius of gyration, $R_g \propto aN^{\nu}$. In spite of such a similarity, the index $\gamma \approx 0.75$ is too high in order to consider the scaling factor $\tilde{a}N^{\gamma}$ as an approximation for the radius of gyration, $R_g \propto aN^{\nu}$ (for the system in question, the maximum possible value of index ν is 0.6). Therefore, the scaling property of Eq. (7.28) seems non-applicable for the explanation of the effect of the molecular weight on the elastic modulus of electrospun nanofibers.

Nevertheless, this contradiction can be resolved by an alternative interpretation of the experimental data. The point is that the index ν in the scaling factor $R_g \propto aN^{\nu}$ can be modified within the accuracy of the experimental data. In other words, the values of fiber radius in the experimental data can be rescaled by the factor $N^{\tilde{\nu}}$, with another index, $\tilde{\nu}$, being higher than ν. Indeed, such a rescaling by the factor N^{γ} with $\gamma \approx 0.75$ results in the fact that all experimental points collapse on one curve (see see Fig. 7.3).

Moreover, the linear fitting of the rescaled experimental data in the double-log coordinates results in the following equation for the approximating curve:

$$E_{\text{rel}} \approx 1 + 35\left(\frac{R}{aN^{0.75}}\right)^{1.5} \tag{7.31}$$

The index value in Eq. (7.31) is 1.5, i.e., slightly less than the index value $\delta \approx 1.6$ in Eq. (7.28). Such a small difference is, in this case, negligible, showing a good agreement of the theoretical model and the experimental data.

Thus, the proposed model do result in the size-dependent mechanical behavior of electrospun polymer nanofibers. However, the real test for the model validity from the physical point of view is a correct estimation of the effect scale. The theoretical dependencies presented in Fig. 7.2 were plotted with the utilizing of physically reasonable values of the system parameters. The comparison of the dependence (7.28), $\hat{E}_{\text{eff}}/E_{\text{bulk}}$ (see Fig. 7.2, solid line), with the experimental values of Young's modulus $E_{\text{eff}}/E_{\text{bulk}}$ for electrospun Nylon-6.6 nanofibers (see Fig. 7.2, black [12], red [34], and blue [93] circles) demonstrates a tolerable agreement. Thus, the proposed model can predict correctly also the scale of the phenomenon in question. And this argument seems of most importance.

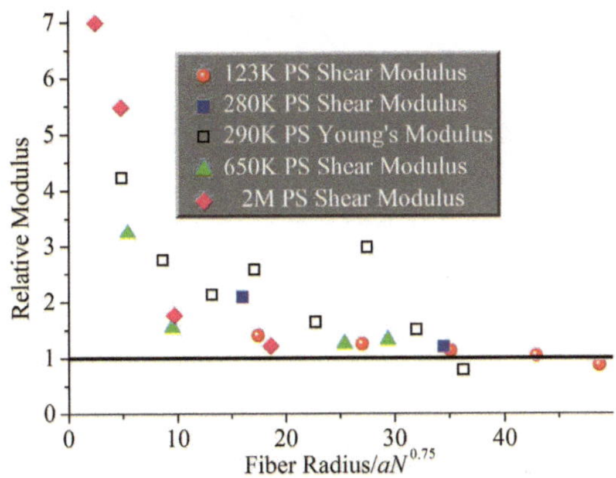

Figure 7.3 Relative moduli of fibers, fabricated of the polymers of various molecular weights, as a function of their radius, rescaled by $\tilde{a}N^{\gamma}$, in place of the scaling factor R_g (the used data are presented in [10]). The fitting of the experimental data results in the function $1 + 35x^{1.5}$, depicted by the dashed line. Reprinted from Ref. [118], Copyright 2013, with permission from John Wiley and Sons.

Note that the discussed model also allows to explain the experimental observation that the molecular weight affects the elastic modulus of electrospun nanofibers demonstrating the scaling dependence (power function). Contrary to the guess of Ji et al. [10], this effect is not related to the scale of the polymer coil and arises due to the fact that the sizes of the ordered regions in the amorphous portion of fiber polymer matrix are being defined by the parameters of polymer system, in particular, by the polymer molecular weight.

7.2 Shifting of Melting Point for Semi-Crystalline Polymer Nanofibers

The experimental results obtained by Arinstein et al. [125] show a decrease in the melting temperature of electrospun microscale fibers (a few microns) with a decrease in their diameter (see Fig. 7.4).

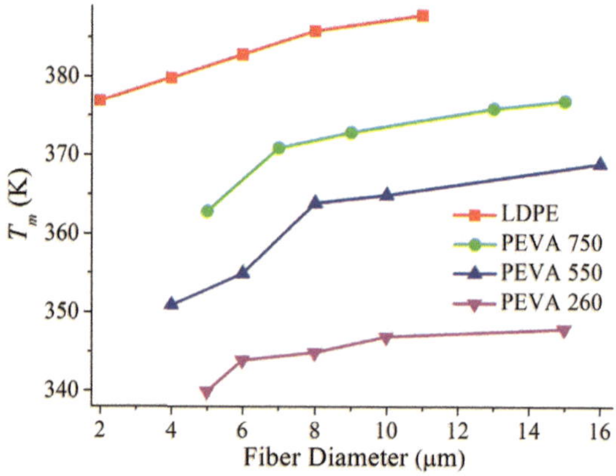

Figure 7.4 Melting temperature versus diameter of LDPE, PEVA 750, PEVA 550, and PEVA 260 nanofibers. The melting temperatures of the nanofibers were obtained using SMFM. The melting temperature of the thickest fibers for each polymer coincided with the bulk melting temperature. Reprinted with permission from Ref. [125], Copyright 2011, IOP Publishing.

Similar behavior was obtained by Liu et al. for PEVA fibers with clay [25]. Wang and Barber have also observed depression of the melting temperature of electrospun polyethylene oxide (PEO) nanofibers [23]. Note that PEO nanofibers have much smaller diameters (below 600 nm) as compared to PEVA and LDPE fibers. Wang and Barber [23] have explained the shift in the melting temperature by adapting the Gibbs–Thompson equation, which is commonly used to describe the shift in the melting temperature of mono-crystals, lamellas, or spherulites having a size of about 10 nm [126]:

$$\frac{T_{\mathrm{m}}(D)}{T_{\mathrm{m}}(\infty)} = \exp\left[-\frac{2S_{\mathrm{vib}}(\infty)}{3R(D/D_{\mathrm{cr}}-1)}\right], \tag{7.32}$$

where $T_m(D)$ and $T_m(\propto)$ are the melting temperatures of nanocrystals and the corresponding bulk value, respectively. R is the ideal gas constant, and S_{vib} (∞)is the vibrational contribution to the overall melting entropy S_m (∞). If the diameter of a fiber is D, the critical diameter at which all the molecules or atoms are at the surface is given by D_{cr}.

However, the applicability of the Gibbs–Thompson model is questionable, since the scale of electrospun polymer fibers demonstrating a melting temperature depression exceeds that of the aforementioned crystalline structures by two orders of magnitude, so an alternative explanation is required. Such a model, as discussed below, was proposed by Arinstein et al. [125].

7.2.1 Theoretical Background

In order to explain the experimentally observed size dependence of the melting temperature of electrospun fibers, it was assumed that the influence of the fabrication stage had a crucial impact on the final thermodynamic properties of these fibers. This is due to the nonequilibrium thermodynamic state created in the polymer matrix during the electrospinning process, when the spinning jet undergoes extreme elongation (the draw ratio of the jet can reach ~ 10^5), stretching the polymer molecules within it. As this occurs, the rapid solvent evaporation fixes the polymer matrix in this stretched, yet nonequilibrium, state. While this process lasts less than 10 ms, the typical relaxation time in a condensed polymer system can extend to several seconds [127]. On the other hand, the evaporation of any residual solvent trapped within the electrospun fibers can also last several seconds, and during this period, the solvent can act as plasticizer [65], allowing the polymer matrix " relax" and reach a state of equilibrium. However, crystallization and/or spherulite formation [25] do not allow for relaxation in the amorphous interlayers between crystals due to the presence of tie molecules. We will focus on these amorphous interlayers in their nonequilibrium state to explain the physical mechanism that produces the observed shift in the melting temperature of the semi-crystalline electrospun fibers.

Let us draw our attention to the jump in entropy, ΔS, in this phase transition. In the case of nanofiber melting, the entropy jump increases because of an additional contribution, δS_{fiber}, related to the nonequilibrium internal structure of the electrospun fiber that results in the observed shift in the melting temperature that results in the observed decrease in the melting temperature. Indeed, the temperature of the melting phase transition is related to jumps in energy and entropy as $T_m = \Delta E/\Delta S$, so in the case of electrospun fibers, their melting temperature is

$$T_m^{\text{fiber}} = \frac{\Delta E}{\Delta S} = \frac{Q}{\Delta S_{\text{bulk}} + \delta S_{\text{fiber}}} < \frac{Q}{\Delta S_{\text{bulk}}}, \tag{7.33}$$

where $Q \equiv \Delta E$ is the specific melting heat.

The internal structure of a semi-crystalline electrospun polymer fiber is characterized by a high degree of orientation of both the polymer crystals and the amorphous part of the polymer matrix [12]. At first glance, the additional jump in entropy can be attributed to this orientation of the polymer crystals, which results in a decrease in the overall entropy of the fiber to below the melting temperature. However, this assumption is rendered invalid by the fact that the melting temperature of ordinary (macro-) fibers is no different than that of bulk in spite of the fact that the crystal orientation in the macro-fibers is also high. It would seem that the crystal orientation does not affect the melting temperature as the crystals inside the fibers are isolated and do not interact with each other.

The influence of the amorphous part of the polymer matrix upon the state of polymer crystals is a very complicated mechanism. On one hand, the interface of the micro-crystals in electrospun fibers is relatively large; therefore, the amorphous part of the polymer matrix may have an effect on the crystals' state. However, in ordinary thick fibers, we have the same situation with no shift in the melting temperature. One can discern only one structural parameter that is different between nano- and thick fibers: the level of chain entanglement. The fact is that the difference in the states of the polymer matrix for bulk and thick fibers and for electrospun fibers is related to the initial fabrication conditions. Formation of electrospun fibers occurs under very fast solvent evaporation; this results in the fixation of a macromolecule entanglement corresponding to the initial concentration of the semi-dilute polymer solution used in the electrospinning process. Also the increase in topological knots in the polymer matrix during jet solidification is hindered by macromolecule disentanglement, which occurs as a result of extreme elongation of the electrospun jet. The formed topological network is not able to "relax" and establish an equilibrium state because of tie chains incorporated into crystals. Indeed, fixation of a chain end results in a fixed topology of polymer matrix. A change in the topology is possible only after a chain end becomes free and that occurs only during melting. After there is melting of the crystals, which had been fixing the topology of the polymer matrix, the concentration of topological knots increases. The entropy that corresponds to the network of topological knots also increases, and this increase results in an additional jump in the entropy, δS_{fiber}.

The physical basis of the aforementioned mechanism can be explained in terms of tie chains as follows: Tie molecules, which try to leave the crystal in order to allow for an increase in the degree of entanglement, weaken the crystal and promote its destruction. As a result, the crystal melting requires less energy and occurs at a lower temperature. In terms of thermodynamics, one can say that the low degree of entanglement of chains in the amorphous part of the polymer matrix does not permit the equilibrium packing, and the system tends to become dense. As a result, a negative additive in the interface pressure between amorphous and crystal polymer occurs, thus causing a decrease in the melting temperature.

Let us estimate the additive in entropy jump, δS_{fiber}, as a function of fiber diameter. Assuming that the system of topological knots is similar to a gas, the entropy for one knot is equal to the logarithm of its concentration, $\ln(c_{\text{kn}})$. The knot concentration is less than the polymer concentration in n_{kn} times (n_{kn} is the mean number of monomers between two neighboring knots along the contour of the polymer chain). Therefore, the additive in entropy jump, δS_{fiber}, is

$$\delta S_{\text{fiber}} = \ln\left(n_{\text{kn}}^{\text{solv}} / n_{\text{kn}}^{\text{bulk}}\right) \tag{7.34}$$

The mean distance between two neighboring knots can be estimated by a scaling law

$$\xi = R_g(\phi_{\text{solv}} / \phi^*)^{\nu/(1-3\nu)}, \tag{7.35}$$

where ϕ_{solv} is the volumetric concentration of the solvent used in the electrospinning process, $\phi = N^{1-3\nu}$ is the volumetric crossover concentration corresponding to the polymer chain overlap, N is the monomer number of polymer chains, $R_g = l_0 N^{\nu}$, $\xi = l_0\left(n_{\text{kn}}^{\text{solv}}\right)^{\nu}$, and index ν is assumed below as $\nu = 0.5$. At the same time, the mean number of monomers between two neighboring knots along the contour of the polymer chain is related to the mean distance between two neighboring knots, ξ, as $n_{\text{kn}}^{\text{solv}} = (\xi / l_0)^2$. Therefore, $n_{\text{kn}}^{\text{solv}}$ is determined by the solvent concentration used in the electrospinning process, according to the following:

$$n_{\text{kn}}^{\text{solv}} = N(\phi_{\text{solv}}/\phi^*)^{-2} \tag{7.36}$$

The diameter of electrospun fibers is dependent on the initial concentration, ϕ_{solv}, of the spinning polymer solution. This experimental pattern can be described well with the help of the following empirical scaling equation [52]:

$$D_{\text{fiber}}/D_0 = (\phi_{\text{solv}}/\phi^*)^\beta = (c_{\text{solv}}/c^*)^\beta, \tag{7.37}$$

where $\beta = 3.1$, and $D_0 = 0.01$ μm; c_{solv} and c^* are the corresponding weight concentrations. The experimental data obtained by Ying et al. [25] for PEVA 260 electrospun fibers result in the following values of parameters: $\beta = 2.99$, and $D_0 = 0.11$ μm (see Fig. 7.5).

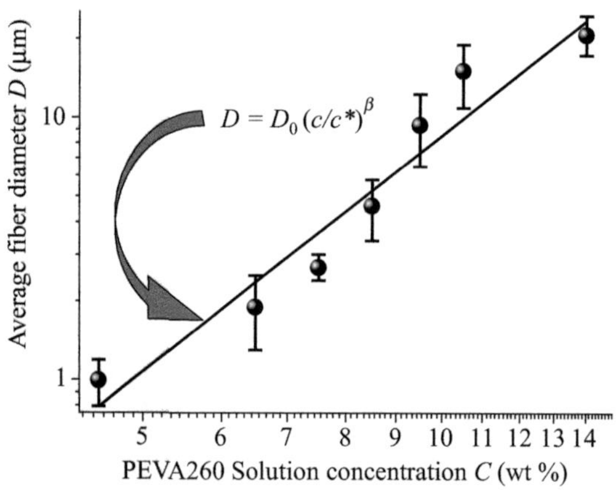

Figure 7.5 Diameter of as-spun nanofibers versus concentration of polymer solution. The fitting was done with the help of the function $D_{\text{fiber}} = D_0(c/c^*)^\beta$ and results in the following parameters: $\beta = 2.99$ and $D_0 = 0.11$ μm ($c^* = 5$ wt% [128]). Reprinted with permission from Ref. [125], Copyright 2011, IOP Publishing.

Substituting Eqs. (7.36) and (7.37) into Eq. (7.34), we get the final equation for the additive in entropy jump, δS_{fiber}, as a function of fiber diameter:

$$\delta S_{\text{fiber}} = \ln\left[\left(N / n_{\text{kn}}^{\text{bulk}}\right)\left(D_0 / D_{\text{fiber}}\right)^{2/\alpha}\right] \tag{7.38}$$

$$= \ln\left(ND_0^{0.67} / n_{\text{kn}}^{\text{bulk}}\right) - 0.67\ln\left(D_{\text{fiber}}\right)$$

Using this equation (7.38), the melting temperature of electrospun fibers as a function of fiber diameter can be written in the following form, which is a linear function of ln (D_{fiber}):

$$\frac{T_{\text{melt}}^{\text{bulk}}}{T_{\text{melt}}^{\text{fiber}}} = 1 + \frac{T_{\text{melt}}^{\text{bulk}}}{Q}\delta S_{\text{fiber}} = A - B\ln\left(D_{\text{fiber}}\right) \tag{7.39}$$

where $A = 1 + \left(T_{\text{melt}}^{\text{bulk}} / Q\right)\ln\left(ND_0^{0.67} / n_{\text{kn}}^{\text{bulk}}\right)$, $B = 0.67T_{\text{melt}}^{\text{bulk}} / Q$, and $T_{\text{melt}}^{\text{bulk}} = Q / \Delta S_{\text{bulk}}$ is the melting temperature of the bulk.

7.2.2 Experimental Results and Discussions

The order of magnitude of parameters contained in Eq. (7.39) can be estimated by the example of PE, whose properties are not too different from the properties of the materials tested. In order to estimate the specific melting heat Q, we have to take into account that it corresponds to the part of the polymer chain between two neighboring topological knots, and this loop consists of $n_{\text{kn}}^{\text{bulk}} \approx 10^2$ monomers. Therefore, $Q = 71 \times 28 \times 10^2 \approx 2 \times 10^5$ J/mol (the specific melting heat of PE is 71 J/g, and the molecular weight of its monomer is 28), and $B = 0.67RT_{\text{melt}}^{\text{bulk}}[\text{K}]/Q \approx 0.011$ (R = 8.314 J/mol.K). Assuming that $N \approx 10^4$ (which corresponds to the typical degree of polymerization of PE), we obtain $A = 1 + \left(RT_{\text{melt}}^{\text{bulk}}[\text{K}]/Q\right)\ln\left(ND_0^{0.67} / n_{\text{kn}}^{\text{bulk}}\right) \approx 1.05$.

Fitting of experimental data to the obtained linear dependence (7.39) is presented in Fig. 7.6. One can observe that both the theoretical dependence (7.39) and the above numerical estimations are in good agreement with the experimental data.

Although the proposed model satisfactorily describes the experimental results, some details of the phenomenon were not taken into account. A more comprehensive model should consider the actual final state of the topological knot system, as well as the effect of tie molecules on the crystalline state in the electrospun fibers. Still, however, the dominant factor of the phenomenon related to the topology of the knot network in the polymer matrix of the electrospun fibers is taken into account and allows one to well capture the general trend of the shift of the melting point.

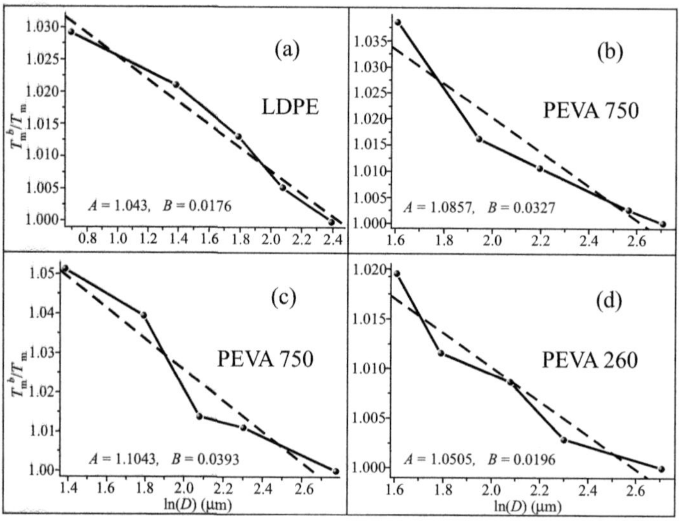

Figure 7.6 Fitting of experimental values of melting temperatures versus diameter by linear equation (46) for LDPE (a), PEVA750 (b), PEVA550 (c), and PEVA 260 (d) as-spun nanofibers. Reprinted with permission from Ref. [125], Copyright 2011, IOP Publishing.

Finally, we can repeat our main idea regarding the explanation of the size-dependent behavior also for thermodynamic properties with the help of surface effects [105, 106]. At first glance, such an explanation seems reasonable. For example, the shift in glass transition temperature of polymer materials being confined into nanopores or nano-slots depends on the type of material used to confine the polymer specimen being tested [129]. However, it turned out that the observed modification in thermodynamic properties of nano-objects also cannot be explained quantitatively by the surface effects, as the estimation of the effect scale is only several nanometers [96, 130], while the above size-dependent behavior can be observed on the scale by two or three orders of magnitude higher (in the discussed experiments, the shift in melting temperature was observed even for electrospun fibers with diameters of several micrometers). The proposed alternative explanation of the shift in the melting temperature of electrospun semi-crystalline polymer fibers is based solely on the aforementioned statement according to which the degree of entanglement (topology) of the polymer matrix does not correspond to an equilibrium state. And just this nonequilibrium, the degree of which is being determined by the electrospinning conditions required for the fabrication of fibers of a given diameter, results in a decrease in the melting temperature.

7.3 Nonequilibrium State of Polymer Matrix Inside Electrospun Nanofibers

One of the key points in the above analysis is the assumption regarding a stretched state of polymer macromolecules inside electrospun nanofibers. Such an assumption is quite reasonable from the physical point of view, as all stages of nanofiber fabrication with the help of electrospinning are accompanied by a high and rapid stretching of the used materials; and such extreme processing conditions result in a nonequilibrium stretched state of nanofiber polymer matrix, which can be " frozen" due to massive solvent evaporation and polymer solidification. And although the collected electrospun nanofibers still contain residual solvent, which slowly evaporates out of the solidified fibers, allowing partial relaxation of stretched polymer molecules to their equilibrium, the final conformations of polymer chains remain nonequilibrium [65]; and this fact (more exactly, assumption) is commonly accepted. Nevertheless, only "theoretical" reasons and arguments are insufficient for serious researchers; and in such a situation, an experimental examination is required in order to validate this assumption.

7.3.1 Orientation of Polymer Macromolecules as a Result of Electrospinning

It is clear that massive elongation of the polymer solution during electrospinning results in the stretching of polymer macromolecules related to elastic deformation mode, and consequently in the formation of nanofibers with a high degree of molecular orientation. Therefore, orientation measurements, as believed, are suitable to demonstrate the stretched state of polymer macromolecules and to quantitatively characterize the degree of this stretching.

One of the possible experimental examination of the internal structure of electrospun polymer nanofibers is related to WAXS. In addition to information regarding crystallinity, this method allows also to analyze anisotropy in tested specimens. The point is that the scattering on the anisotropic objects results in an anisotropic distribution of diffraction intensity. Typical two-dimensional WAXS patterns demonstrating such an anisotropy are presented in Fig. 7.7 (these images were obtained at the X-ray testing of the electrospun Nylon-6,6 nanofibers with different diameters).

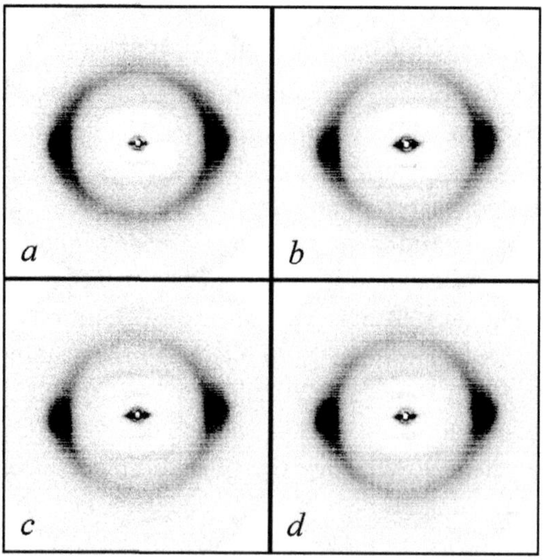

Figure 7.7 Two-dimensional WAXS patterns of the electrospun PA66 nanofibers having an average diameter of (a) 900 nm, (b) 600 nm, (c) 300 nm, and (d) 100 nm. Reprinted from Ref. [12, Supplementary Section I], Copyright 2007, Nature Publishing Group.

Decomposition of a radial WAXS signal of nanofibers into a set of Gaussian peaks allows one to separate out amorphous and crystalline contributions, calculating the content of each component separately in accordance with the amplitude of each of the peaks. For example, decomposition of the equatorial WAXS signal consisting of one amorphous and two crystalline peaks is shown in Fig. 7.8. (Note that decomposition of multi-peak signal into component peaks is a very hard problem; and in the cases of significant overlapping of individual peaks, we can deal with so-called ill-posed inverse problem unique solution of which is impossible.)

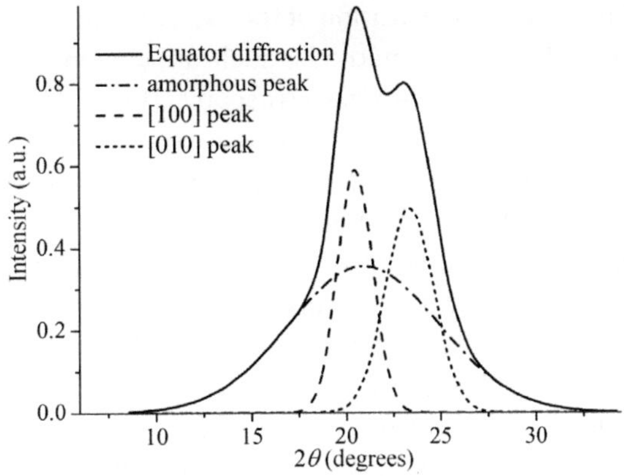

Figure 7.8 Decomposition of the equatorial WAXS signal of nanofibers having an average diameter $D \approx 100$ nm. Reprinted from Ref. [12, Supplementary Section I], Copyright 2007, Nature Publishing Group.

The knowledge of the amplitudes of individual peaks obtained as a result of decomposition of radial SAXS signals from different directions allows one to plot the azimuthal distributions of the WAXS signal for each peak separately, which in turn allows one to estimate the anisotropy of the examined system.

In the case of narrow azimuthal peaks, the anisotropy level f can be calculated as follows:

$$f = \frac{90^\circ - \Delta\varphi}{90^\circ}, \tag{7.40}$$

where $\Delta\varphi$ is the half-width of an azimuthal peak at half of its maximum height.

In the case of broad azimuthal peaks (usually, such broad peaks correspond to the relatively low oriented amorphous system component), the anisotropy level, characterized by the order parameter η, is calculated as follows:

$$\eta = \frac{1}{2}\left\langle 3\cos^2\theta - 1\right\rangle = \frac{1}{2}\int_0^{\pi/2}\left(3\cos^2\theta - 1\right)P(\theta)\sin\theta d\theta, \tag{7.41}$$

where $P(\theta)$ is the normalized azimuthal amorphous peak, which appears in an oriented system in place of homogeneous amorphous halo.

Note that X-ray analysis allows one to determine only total level of orientational ordering, its mean value, and when we are interested in the characterization of the local orientational ordering, we have to utilize other methods, such as transmission electron microscopy [131], infrared spectroscopy (IR) [132], polarized Raman spectroscopy [133, 134], etc. (see also, excellent review of Richard-Lacroix and Pellerin "Molecular Orientation in Electrospun Fibers: From Mats to Single Fibers" [135]).

Thus, the examination, based on confocal Raman spectroscopy, was presented by Richard-Lacroix where individual electrospun nanofibers of poly(ethylene terephthalate) (PET) aligned with their long axis along the Z-direction were tested [134]. It turned out that a very broad distribution of molecular orientation and structure within the samples was observed: Some individual fibers are completely isotropic and amorphous, while others present an order parameter η, calculated according to Eq. (7.41), as large as 0.75.

Figure 7.9 presents an example of a full set of parallel- (XX and ZZ) and cross-polarized (XZ and ZX) Raman spectra for a ~ 500 nm electrospun PET fiber in the 1575–1775 cm^{-1} spectral region.

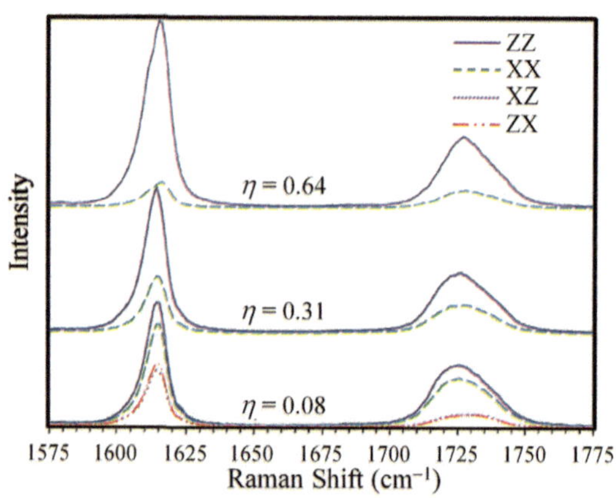

Figure 7.9 Polarized Raman spectra for three representative individual electrospun PET fibers in the 1575–1775 cm^{-1} spectral region. For clarity, the cross-polarized spectra (XZ and ZX) are only shown for one fiber. Reprinted with permission from Ref. [134]. Copyright 2012, American Chemical Society.

The molecular orientation in these nanofibers was quantified using the ratios of intensities of the bands with different polarization. Note that the width and the position of the 1616 cm^{-1} band do not depend on the orientation and/or crystallization. In contrast, the 1725 cm^{-1} band shifts and narrows with increasing crystallinity.

Figure 7.10 shows the polarized Raman spectra of the same three samples in the 675–1175 cm^{-1} spectral region. The bands are considerably weaker (as compared to the 1575–1775 cm^{-1} spectral region), especially in the cross-polarized spectra since most of them have a small depolarization ratio.

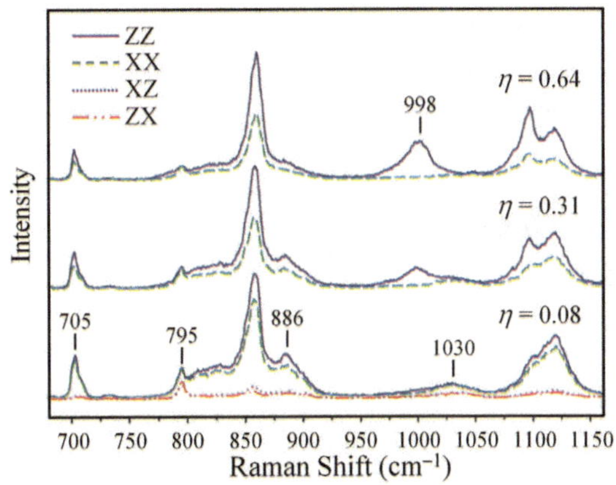

Figure 7.10 Polarized Raman spectra of single electrospun PET fibers in the 675–1175 cm^{-1} spectral region. For clarity, the cross-polarized spectra (XZ and ZX) are shown only for the bottom spectrum. Reprinted with permission from Ref. [134]. Copyright 2012, American Chemical Society.

These spectra strongly support the quantitative estimation of orientation in PET nanofibers (for details, see [134]).

Now we would like to draw our attention to some peculiarities in *trans-gauche* isomerism behavior. The 998 cm^{-1} band, which is associated with O–CH_2 and C–C stretching of the ethylene glycol unit in the *trans* conformation, appears only in the ZZ spectra of oriented nanofibers and its intensity increases with orientation. A similar but more complicated behavior is observed for the 1096 cm^{-1} band, which is also associated with the *trans* conformation. This is in agreement with the *gauche*-to-*trans* rotational isomerization of the ethylene glycol units upon orientation. These results indicate that the population of *trans* conformer increases with the overall orientation of the polymer chains and that their degree of orientation is extremely high in all cases. In contrast, the bands associated with *gauche* conformers show the opposite behavior. In particular, the 1030 cm^{-1} band does not show any polarization dependence and only appears in the spectra of weakly oriented fibers, while the 886 cm^{-1} band shows a drastic decrease in intensity in both parallel-polarized spectra upon orientation. These results suggest that the *gauche* conformers are essentially isotropic and that their fraction decreases with the overall orientation of the system.

Figure 7.11 shows the evolution of the *trans* and *gauche* fractions determined independently as a function of the order parameter η, for a series of nanofibers covering the complete range of obtained orientation levels.

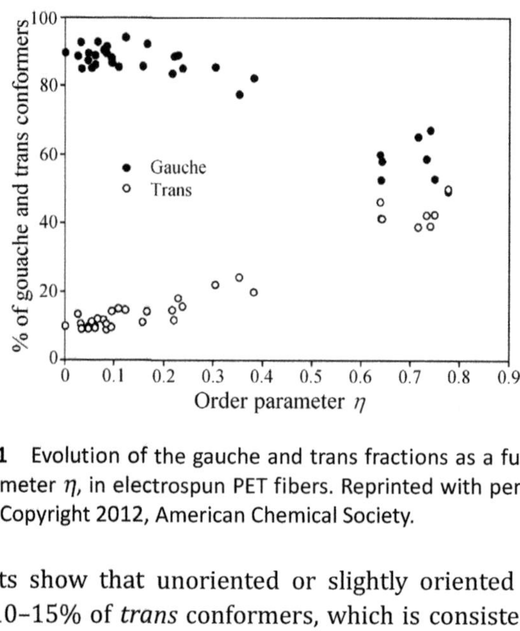

Figure 7.11 Evolution of the gauche and trans fractions as a function of the order parameter η, in electrospun PET fibers. Reprinted with permission from Ref. [134]. Copyright 2012, American Chemical Society.

Results show that unoriented or slightly oriented nanofibers contain 10–15% of *trans* conformers, which is consistent with the conformation distribution of totally amorphous and isotropic PET samples. The population of *trans* conformers gradually increases with orientation increase and reaches 40–50% for the nanofibers with the highest η values. At the same time, the population of *gauche* conformers follows the opposite trend.

Thus, we can see that the electrospun nanofibers do demonstrate a high level of orientation ordering; and this orientation ordering is nowadays being associated by many researchers with a stretching of polymer macromolecules in electrospun nanofibers. However, orientation measurement as a means of estimating the polymer stretching in electrospun nanofibers contains an internal contradiction. The stretched macromolecules should demonstrate some orientation, whereas the orientational ordering can be observed even in non-stretched systems, for example, in liquid crystals. In spite of that, there are no doubts that the molecular orientation, observed in the aforementioned measurements, is related to polymer stretching. Nevertheless, the direct experiment demonstrating the stretching electrospun polymer nanofiber can be very helpful.

7.3.2 Stretching of Polymer Matrix of Electrospun Nanofibers

It turned out that such a direct demonstration of a stretching in electrospun polymer nanofibers is possible. Thus recently, Alhazov et al. have shown that thermo-mechanical behavior of electrospun thermoplastic polyurethane (TPU) nanofibers dramatically differs from that of cast TPU films [136]. Upon heating at a constant rate (1°C/min or 5°C/min), an electrospun mat of nanofibers undergoes significant contraction (up to tens of percent), whereas films expand as expected (see Fig. 7.12).

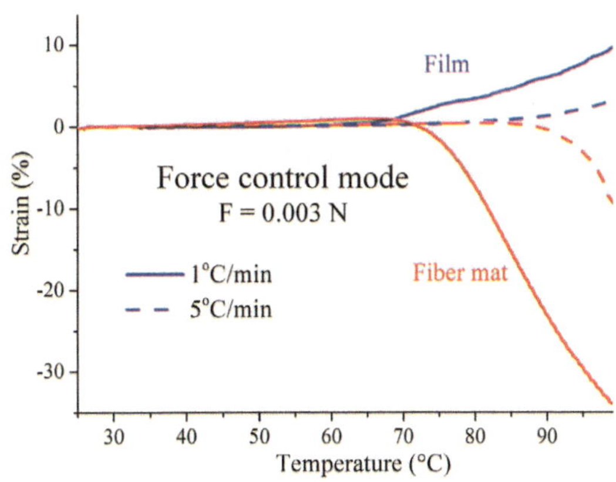

Figure 7.12 Strain versus temperature response of TPU-based cast film and electrospun fibers with diameter in the range between 500 and 800 nm. The experiment was carried in dynamic mechanical analysis machine (DMA) at force-control mode. (TPU - Krystalflex PE-399, Huntsman, USA). Reprinted from Ref. [136], Copyright 2013, with permission from Elsevier.

As the internal fiber network structure may affect the contraction kinetics and, possibly, the final state of the fiber mat, an examination of a single electrospun nanofiber is necessary in order to understand the features of the phenomenon in question. For this goal, two groups of electrospun nanofibers fabricated by applying two different electrostatic fields (1.14 and 1.81 kV/cm) were tested by Alhazov et al. [137]. Typical experimental results of the thermo-mechanical testing of single electrospun nanofibers upon heating, obtained with the help of a custom-made experimental system, are shown in Fig. 7.13.

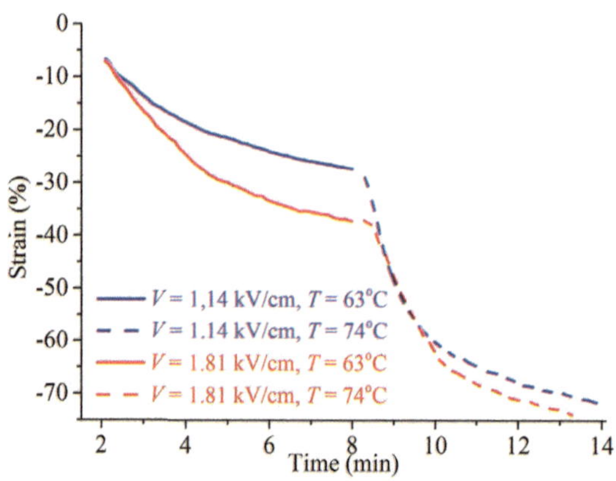

Figure 7.13 Experimental observations of single nanofiber contraction versus time under isothermal condition for fibers spun at 1.14 kV/cm and at 1.81 kV/cm. During the first 8 min, the ambient temperature was 63°C, and was then raised to 74°C. Reprinted from Ref. [137], Copyright 2015, with permission from John Wiley and Sons.

One can see that at 63°C, nanofibers fabricated under an electrostatic strength of 1.14 kV/cm contracted up to a strain value of ~ 28%, whereas the contraction of the fibers fabricated under 1.81 kV/cm reached ~ 37%. These findings fall in line with previous reports of increased stretching of the polymer matrix in electrospun fibers following an increase in the electrostatic field with the same flow rate. Upon an additional temperature increase of up to 74°C, both types of fibers demonstrated an additional contraction of up to ~ 70% their original size. Note that fiber length in both types of fibers initially contracted by ~ 7% relative to the length measured immediately after fabrication, which is apparently due to relaxation of the residual stress remaining as a result of spinning.

A possible explanation for shortening of the TPU nanofibers is relaxation of the stretched polymer matrix in electrospun nanofibers. Indeed, as was mentioned earlier, the electrospinning process results in the stretching of the polymer matrix inside the electrospun nanofibers, which can be preserved upon rapid solvent evaporation. And in order to prove this statement by real experiment, the thermal response of a pre-stretched TPU-system under the same (or similar) thermo-mechanical conditions and pre-history (processing conditions) should be analyzed. To this end, the classical thermo-mechanical cyclic test for a film, simulating the stretching of the polymer chains and fast evaporation during the electrospinning process, was executed [137]. A film was heated up to a high temperature (~ 90°C) and kept at this temperature for 10 min. Thereafter, the sample was stretched up to strain of about 100%. After deformation, the sample was kept under isothermal conditions in a strain-control mode for 10 min, and, finally, the sample was cooled to room temperature. Then, the test regime was switched to a force-control mode (the stress was 0.003 N), and after exposure at room temperature for 10 min, the stretched sample was heated up to the same high temperature (~ 90°C). The experimental results are shown in Fig. 7.14.

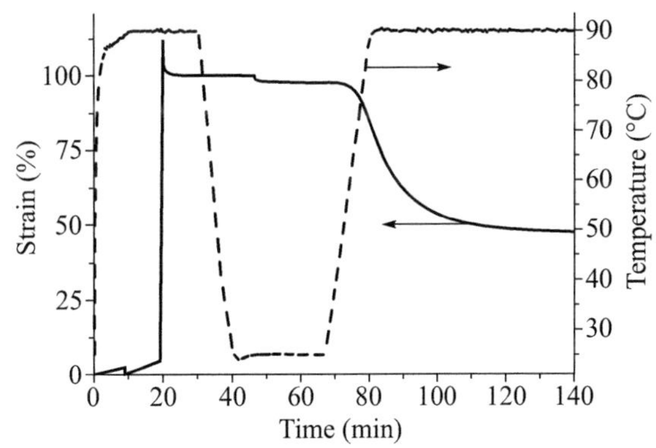

Figure 7.14 Typical thermo-mechanical cycle test results for a TPU film: strain (solid line) and temperature (dashed line) versus time. The film was maintained for 10 min before deformation at 90°C, then for 10 min after deformation at 90°C, and for 10 min after cooling to 25°C. Reprinted from Ref. [137], Copyright 2015, with permission from John Wiley and Sons.

One can see that the TPU cast films demonstrate a behavior similar to the electrospun nanofibers after subjection to thermo-mechanical processing steps resulting in the formation of a nonequilibrium stretched state of the polymer matrix; and this observation is the direct experimental proof of the fact that the state of polymer in electrospun nanofibers is stretched.

7.3.3 Orientation versus Stretching under Confinement

Completing our discussion regarding stretching and orientation of polymer macromolecules inside electrospun nanofibers, it is suitable to mention the results obtained by Deng et al. at the analysis of mechanical properties of glassy polymers under cylindrical confinement via molecular dynamics simulations [138].

In this study, two types of systems were examined: The first group of cylindrical fibers was obtained by a decrease in diameter of confinement cylinder down to different values with no change in the pressure in the direction along the fiber axis, whereas such cylindrical fibers with different diameters of the second group were obtained by the force-controlled stretching with the help of a force linearly increasing along the fiber axis. It turned out that Young's moduli of the fibers of both groups sharply increased below a crossover radius, i.e., both groups demonstrated size-dependent behavior (see Fig. 7.15).

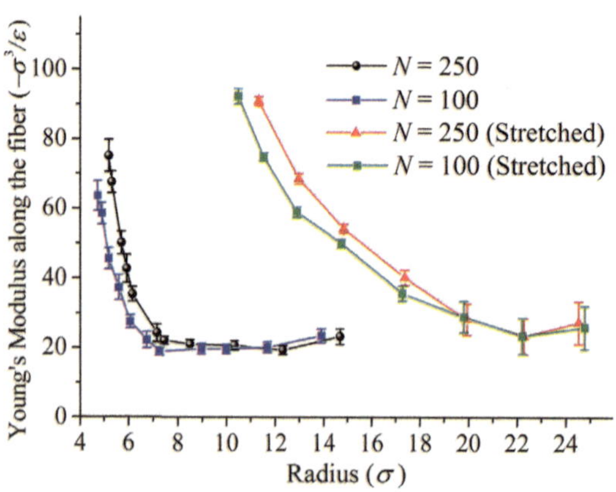

Figure 7.15 Young's modulus versus fiber radius below T_g for system under cylinder confinement (two left curves) and for system with force-controlled stretching pretreatment (two right curves). Reprinted from Ref. [138], Copyright 2017, with permission from John Wiley and Sons.

The difference between groups is only in the crossover scale: The crossover radius for the fibers obtained by a shrinking is of about 7 σ, whereas the crossover radius for the fibers obtained by stretching is larger, of about 23 σ (where σ is the scale of monomer, which can be assumed as 0.2–0.5 nm).

In addition, it turned out that the increase in the modulus is accompanied by modification of the polymer structure, reflected, in particular, by enhancement in the chain alignment. In doing so, this alignment enhancement (more exactly, orientational ordering) below crossover radius was observed in both fiber groups. If such an orientational ordering in the stretched systems is quite expected, the reason for such effect in shrunken system requires an explanation. Indeed, in the case of the first fiber group, no orienting forces were used during the fiber preparation, so no orientational ordering should appear. Nevertheless, the shrinking of the fiber down to the scale comparable to the scale of the equilibrium macromolecule conformation corresponding to Gaussian ball should result in the initiation of an anisotropy in macromolecule conformation. The point is that the isotropic Gaussian ball appears as a result of the averaging over all possible conformations of the test macromolecule, including the stretched ones: both along the fiber axis and in the perpendicular directions. If the fiber diameter is small enough (smaller than the scale of a stretched macromolecule), the last stretched conformations are impossible as the macromolecule cannot go out of the fiber. As a result, the stretched conformations along the fiber axis are not compensated by corresponding conformations in the perpendicular directions; and the anisotropy in macromolecular structure of such thin fibers appears. With a decrease in the fiber diameter, the aforementioned unbalancing between stretched conformation in different directions increases; therefore, the orientational ordering should increase when fiber diameter decreases. In other words, there exists a dependence of the order parameter describing the level of the system orientational ordering versus fiber diameter. Using this dependence, we can re-plot the data presented in Fig. 7.15, as the fiber's Young's modulus versus order parameter η of the macromolecular system inside these fibers. This dependence is presented in Fig. 1 where we can observe a surprising effect: all data drop on one universal curve.

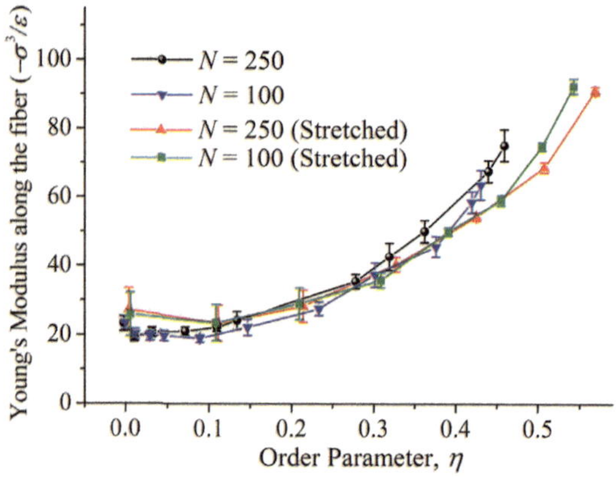

Figure 7.16 Young's modulus versus order parameter η, below T_g. The increase in Young's modulus is mainly caused by the increase in the overall chain alignment. Reprinted from Ref. [138], Copyright 2017, with permission from John Wiley and Sons.

Such a behavior means that in the case of confinement of individual macromolecules, the reinforcement mechanism is related to the increase in macromolecule orientational ordering, i.e., the question "what is dominant in the fiber reinforcement mechanism: macromolecular stretching or their orientational ordering?" is of no physical importance.

It should be noted that the fiber reinforcement can also occur without increase in the orientational ordering of macromolecules. Indeed, as was shown by Arinstein et al. [12], only a mild, monotonous increase was observed in the orientation of Nylon-6.6 nanofibers as a function of their diameter, whereas Young's modulus of these nanofibers increases dramatically (see Fig. 6.7 in Section 6.4 "Confinement concept" and Fig. 2.8 in Section 2.2.1 "Mechanical properties"). Thus, we can conclude that the physical mechanism that provides the size-dependent behavior of electrospun polymer nanofiber (in particular, their reinforcement) is related not only to the orientational ordering of macromolecules.

The result of the aforementioned computer simulations allows us to formulate one more conclusion. The obtained crossover scale that can be estimated as 5–10 nm is two orders of magnitude lower than the crossover scale observed experimentally. That means the size-dependent behavior of electrospun polymer nanofibers cannot be explained by the confinement of the individual macromolecules, and more complicated confinement mechanisms should be considered.

Final Remarks

The goal of the book was to present and discuss from the physicist point of view the scientific challenges related to the unique mechanical properties and unusual thermodynamic behavior of electrospun polymer nanofibers. Briefly, it can be concluded that

- Due to significant stretching of polymer macromolecules, which occurs during the electrospinning process, a nonequilibrium state of the polymer matrix within electrospun nanofibers is formed. Although this stretched state of polymer matrix can partially relax during final solvent evaporation, the final state of polymer matrix within electrospun nanofibers remains a nonequilibrium one.
- Supermolecular structures arising in nanofiber polymer matrix generate an additional scale, the occurrence of which should be necessarily taken into account when analyzing nano-objects' properties.
- Confinement of these supermolecular structures, which is caused by the shrinking of the transversal nanofiber size, plays a dominant role in the modification of mechanical and thermodynamic properties of nanofibers.

Note that the possibility to generalize the aforementioned ideas for adaptation to other nano-objects requires additional analysis, due to the fact that the final state of the polymer matrix of electrospun nanofibers is significantly affected by nanofiber fabrication conditions.

Finally, taking into account a great number of presented experimental data and the theoretical models based on these data, as well as all the aforementioned discussions, the question "Does small size matter?" can be answered affirmatively without any doubts: "Yes, it does!"

Figure F.1 For electrospun polymer nanofibers, small size does matter, especially, in nonequilibrium conditions.

Bibliography

1. Ch. V. Boys. *Nature* **42**, 604 (1890).
2. J. F. Cooley. GB Patent #06385, May 19, 1900; US Patent #692631, February 4, 1902.
3. E. F. Burton, W. B. Wiegand. *Philosophical Magazine* **23** (133), 148 (1912).
4. J. Zeleny. *Physical Review* **3** (2), 69 (1914).
5. F. Anton. US Patent #1975504, October 2, 1934.
6. Y. Filatov, A. Budyka, and V. Kirichenko. *Electrospinning of Micro- and Nanofiber: Fundamentals and Applications in Separation and Filtration Processes.* New York: Begell House, Inc; 2007.
7. G. Taylor. *Proceedings of the Royal Society A* **313**, 453 (1969).
8. D. H. Reneker and I. Chun. *Nanotechnology* **7** (3), 216 (1996).
9. A. L. Yarin, B. Pourdeyhimi, and S. Ramakrishna. *Fundamentals and Applications of Micro-and Nanofibers.* Cambridge University Press, Cambridge. 2014.
10. Y. Ji, C. Li, G. Wang, J. Koo, S. Ge, B. Li, J. Jiang, B. Herzberg, T. Klein, S. Chen, J. C. Sokolov, and M. H. Rafailovich. *Europhysics Letters* **84**, 56002 (2008).
11. S. Cuenot, Ch. Frétigny, S. Demoustier-Champagne, and B. Nysten. *Physical Review B* **69**, 165410 (2004).
12. A. Arinstein, M. Burman, O. Gendelman, and E. Zussman. *Nature Nanotechnology* **2**, 59 (2007).
13. H. Schonherr and C. W. Frank. *Macromolecules* **36**, 1188 (2003).
14. S. Cuenot, S. Demoustier-Champagne, and B. Nysten. *Physical Review Letters* **85** (8), 1690 (2000).
15. E. P. S. Tan and C. T. Lim. *Applied Physics Letters* **87**, 123106 (2005).

16. M. K. Shin, S. I. Kim, S. J. Kim, H. Lee, and G.M. Spinks. *Applied Physics Letters* **89**, 231929 (2006).

17. Y. Ji, B. Li, Sh. Ge, J. C. Sokolov, and M. H. Rafailovich. *Langmuir* **22**, 1321 (2006).

18. Ph. A. Yuya, Y. Wen, J. A. Turner, Yu. A. Dzenis, and Zh. Li. *Applied Physics Letters* **90**, 111909 (2007).

19. C. T. Lim, E. P. S. Tan, and S. Y. Ng. *Applied Physics Letters* **92**, 141908 (2008).

20. S.-C. Wong, A. Baji, and S. Leng. *Polymer* **49**, 4713 (2008).

21. Q. Fu, Yu Jin, X. Song, J. Gao, X. Han, X. Jiang, Q. Zhao, and D. Yu. *Nanotechnology* **21**, 095703(2010).

22. Y. Liu, Sh. Chen, E. Zussman, Ch. S. Korach, W. Zhao, and M. Rafailovich. *Macromolecules* **44**, 4439 (2011).

23. W. Wang and A. H. Barber. *Nanotechnology*, **21**, 225701 (2010).

24. W. Wang and A. H. Barber. *Journal of Polymer Science, Part B: Polymer Physics* **50**, 546 (2012).

25. Y. Liu, Ch. Li, Sh. Chen, E. Wachtel, T. Koga, J. C. Sokolov, and M. H. Rafailovich. *Journal of Polymer Science, Part B: Polymer Physics* **47**, 2501 (2009).

26. M. Bognitzki, W. Czado, Th. Frese, A. Schaper, M. Hellwig, M. Steinhart, A. Greiner, and J. H. Wendorff. *Advanced Materials* **13**, 70 (2001).

27. S. R. Ge, Y. Pu, W. Zhang, J. S. Sokolov, and M. H. Rafailovich, *Physical Review Letters* **85**, 2340 (2000).

28. W. Wang, A. J. Bushby, and A. H. Barber. *Applied Physics Letters* **93**, 201907 (2008).

29. E. Zussman, M. Burman, A. Yarin, R. Khalfin, and Y. Cohen. *Journal of Polymer Science, Part B: Polymer Physics* **44**, 1482 (2006).

30. E. P. S. Tan, S. Y. Ng, and C. T. Lim. *Biomaterials* **26**, 1453 (2005).

31. E. P. S. Tan, C. N. Goh, C. H. Sow, and C. T. Lim. *Applied Physics Letters* **86**, 1453 (2005).

32. P. A. Yuya, Y. Wen, J. A. Turner, and Y. A. Dzenis. *Applied Physics Letters* **90**, 111909 (2007).

33. A. Arinstein. *Physical Review E* **79**, 056609 (2009).

34. M. Burman, A. Arinstein, and E. Zussman. *Applied Physics Letters* **93**, 193118 (2008).

35. G. I. Taylor. *Proceedings of the Royal Society A* **280**, 383 (1964).

36. G. I. Taylor. *Proceedings of the Royal Society A* **291**, 159 (1966).

37. G. I. Taylor. *Proceedings of the Royal Society A* **313**, 453 (1969).
38. A. F. Spivak, Y. A. Dzenis, and D. H. Reneker. *Mechanics Research Communications* **27**, 37 (2000).
39. J. Doshi and D.H. Reneker. *Journal of Electrostatics* **35**, 151 (1995).
40. M. M. Hohman, M. Shin, G. Rutledge, and M. P. Brenner. *Physics of Fluids* **13**, 2201 (2001).
41. Y. M. Shin, M. M. Hohman, M. P. Brenner, and G. C. Rutledge. *Polymer* **42**, 9955 (2001).
42. D. H. Reneker, A. L. Yarin, H. Fong, and S. Koombhongse. *Journal of Applied Physics* **87**, 4531 (2000).
43. Zh.-M. Huanga, Y.-Z. Zhangb, M. Kotakic, and S. Ramakrishna. *Composites Science and Technology* **63**, 2223 (2003).
44. D. Li and Y. Xia. *Advanced Materials* **16**, 1151 (2004).
45. D. H. Reneker, A. L. Yarin, E. Zussman, and H. Xu. *Advances in Applied Mechanics* **41**, 43 (2007).
46. D. H. Reneker and A. L. Yarin. *Polymer* **49**, 2387 (2008).
47. P. P. Bhat, S. Appathurai, M. T. Harris, M. Pasquali, G. H. McKinley, and O. A. Basaran. *Nature Physics* **6**, 625 (2010).
48. M. Johnson, R. D. Kamm, L. W. Ho, A. Shapiro, and T. J. Pedleyz. *Journal of Fluid Mechanics* **233**, 141 (1991).
49. M. Goldin, J. Yerushalmi, R. Pfeffer, and R. Shinnar. *Journal of Fluid Mechanics* **38**, 689 (1969).
50. V. M. Entov and A. L. Yarin. *Fluid Dynamics* **19**, 21 (1984).
51. N. Ashgriz (ed.) *Handbook of Atomization and Sprays. Theory and Applications.* ISBN 978-1-4419-7263-7. Springer Science & Business Media, LLC, 2011.
52. P. Gupta, C. Elkins, T. E. Long, and G. L. Wilkes. *Polymer* **46**, 4799 (2005).
53. S. L. Shenoy, W. Douglas Batesa, H. L. Frischb, and G. E. Wnek. *Polymer* **46**, 3372 (2005).
54. J. Ren, O. Urakawa, and K. Adachi. *Polymer* **44**, 847, (2003).
55. A. Vega-Gonzalez, P. Subra-Paternault, F. Nudda, and B. Marongiu. *Proceedings of 10th European Meeting on Supercritical Fluids: Reactions, Materials and Natural Products Processing*, Strasbourg/Colmar, France, December12–14, 2005, T19.
56. M. M. Demir, I. Yilgor, E. Yilgor, and B. Erman. *Polymer* **43**, 3303 (2002).
57. S. N. Reznik and E. Zussman. *Physical Review E* **81**, 026313 (2009).
58. T. Han, A. L. Yarin, and D. H. Reneker. *Polymer* **49** (6), 1651 (2008).

59. L. M. Bellan, H. G. Craighead, and J. P. Hinestroza. *Journal of Applied Physics* **102**, 094308 (2007).

60. S. Koombhongse, W. X. Liu, and D. H. Reneker. *Journal of Polymer Science, Part B: Polymer Physics* **39**, 2598 (2001).

61. A. J. Guenthner, S. Khomobhongse, W. Liu, P. Dayal, D. H. Reneker, and T. Kyu. *Macromolecular Theory and Simulations* **15**, 87 (2006).

62. X.-F. Wu, Y. Salkovskiy, and Y. A. Dzenis. *Applied Physics Letters* **98**, 223108 (2011).

63. S. Tripatanasuwan, Z. Zhong, and D. H. Reneker. *Polymer* **48**, 5742, (2007).

64. P. Dayal and T. Kyu. *Physics of Fluids* **19**, 107106 (2007).

65. A. Arinstein and E. Zussman. *Physical Review E* **76**, 056303 (2007).

66. A. Arinstein, R. Avrahami, and E. Zussman. *Journal of Physics D: Applied Physics* **42**, 015507 (2009).

67. Y. Peng, P. Y. Wu, and Y. L. Yang. *Journal of Chemical Physics* **119**, 8075–8079 (2003).

68. L. Larrondo and R. S. J. Manley. *Journal of Polymer Science, Part B: Polymer Physics* **19**, 921 (1981).

69. J. S. Stephens, S. Frisk, S. Megelski, J. F. Rabolt, and D. B. Chase. *Applied Spectroscopy* **55**, 1287 (2001).

70. P. Dayal, J. Liu, S. Kumar, and T. Kyu. *Macromolecules* **40**, 7689 (2007).

71. C. L. Casper, J. S. Stephens, N. G. Tassi, D. B. Chase, and J. F. Rabolt. *Macromolecules* **37**, 573 (2004).

72. P. Dayal and T. Kyu. *Physics of Fluids* **19**, 107106 (2007).

73. A. Arinstein and E. Zussman. *Journal of Polymer Science, Part B: Polymer Physics* **49**, 691 (2011).

74. I. Greenfeld, A. Arinstein, K. Fezzaa, M. H. Rafailovich, and E. Zussman. *Physical Review E* **84**, 041806 (2011).

75. V. G. Kulichikhin, A. Ya. Malkin, A. V. Semakov, I. Yu. Skvortsov, and A. Arinstein. *European Physical Journal E* **37**, 10 (2014).

76. P. G. de Gennes. *Journal of Chemical Physics* **60**, 5030 (1974).

77. N. K. Balabaev, I. P. Borodin, T. I. Borodina, and T. N. Khazanovich. *Polymer Science Series A* **52**, 655 (2010).

78. Y. Wang, X. Liu, K.-S. Im, W.-K. Lee, J. Wang, K. Fezzaa, D. L. S. Hung, and J. R. Winkelman. *Nature Physics* **4**, 305 (2008).

79. R. J. Roe. *Methods of X-ray and Neutron Scattering in Polymer Science*. Oxford University Press, New-York, 2000.

80. R. Sattler, C. Wagner, and J. Eggers. *Physical Review Letter* **100**, 164502 (2008).

81. R. Sattler, S. Gier, J. Eggers, and C. Wagner. *Physics of Fluids* **24**, 023101 (2012).

82. S. Ya. Frenkel', V. G. Baranov, N. G. Bel'nikevich, and Yu. N. Panov. *Vysokomolekulyarnye soedineniya, Letter to Editor* **6**, 1917 (1964) (in Russian).

83. G. V. Vinogradov, A. Ya. Malkin, N. K. Blinova, S. I. Sergeyenkov, M. P. Zabugina, L. V. Titkova, Yu. G. Yanovsky, and V. G. Shalganova. *European Polymer Journal* **9**, 1231 (1973).

84. J. A. Odell, A. Keller, and M. J. Miles. *Polymer* **26**, 1219 (1985).

85. L. A. Archer. *Journal of Rheology* **43**, 1617 (1999).

86. R. N. Haward and R. J. Young (ed.) *The Physics of Glassy Polymers*, 2nd edition. Chapman & Hall, London, 1997, p. 213.

87. E. Zussman, D. Rittel, and A. L. Yarin. *Applied Physics Letter* **82**, 3958 (2003).

88. T. Yoshioka, R. Dersch, A. Greiner, M. Tsuji, and A. K. Schaper. *Macromolecular Materials and Engineering* **295**, 1082 (2010).

89. C. J. Luo, E. Stride, S. Stoyanov, E. Pelan, and M. Edirisinghe. *Journal of Polymer Research* **18**, 2515 (2011).

90. I. Greenfeld and E. Zussman. *Journal of Polymer Science, Part B: Polymer Physics* **51**, 1377 (2013).

91. A. Ya.Malkin, A. V. Semakov, and V. G. Kulichikhin. *Rheological Acta* **50**, 485 (2011).

92. A. Arinstein. *8th European Nonlinear Dynamics Conference, ENOC 2014*, July 6–11, 2014, Vienna, Austria.

93. M. Burman, A. Arinstein, and E. Zussman. *Europhysics Letters* **96**, 16006 (2011).

94. U. Stachewicz and A. H. Barber. *Langmuir* **27**, 3024 (2011).

95. J. He and C. M. Lilley. *Nano Letters* **8**, 1798 (2008).

96. H. S. Park, W. Cai, H. D. Espinosa, and H. Huang. *MRS Bulletin* **34**, 178 (2009).

97. R. E. Miller and V. B. Shenoy. *Nanotechnology* **11**, 139 (2000).

98. S. Curgul, K. J. Van Vliet, and G. C. Rutledge. *Macromolecules* **40**, 8483 (2007).

99. O. Benaicha, S. Abdelmoula, S. Ben Nasrallah, F. Sakli, and C. Caze. Book of Abstracts of the Fiber Society *2003 Spring Symposium on*

Advanced flexible materials and structures: Engineering with fibers, Loughborough University, UK, June 30 – July 2, 2003.

100. L. Sun, R. P. S. Han, J. Wang, and C. T. Lim. *Nanotechnology* **19**, 455706 (2008).
101. A. Mayes. *Macromolecules* **27**, 3114 (1994).
102. K. Tanaka, A. Taura, S.-R. Ge, A. Takahora, and T. Kajiyama. *Macromolecules* **29**, 3040 (1996).
103. T. Kajiyama, K. Tanaka, and A. Takahora. *Macromolecules* **30**, 280 (1997).
104. P. G. De Gennes. *European Physical Journal E* **2**, 201 (2000).
105. P. Sharma, S. Ganti, and N. Bhate. *Applied Physics Letters* **82**, 535 (2003).
106. R. Dingreville, J. M. Qu, and M. Cherkaoui. *Journal of the Mechanics and Physics of Solids* **53**, 1827 (2005).
107. U. Stachewicz, R. J. Bailey, W. Wang, and A. H. Barber. *Polymer* **53**, 5132, (2012).
108. A. E. Arinstein. *Journal of Experimental and Theoretical Physics* **91**, 206 (2000).
109. S. M. Mezikovskii, A. E. Arinstein, and R. Ja. Deberdeev. *Oligomeric State of Substance*. Nova Science Publishers, New York, 2009.
110. P. G. de Gennes. *Scaling Concepts in the Physics of Polymers*. Cornell University Press, Ithaca, 1979.
111. L. Onsager. *Annals of the New York Academy of Sciences* **51**, 627 (1949).
112. A. R. Khokhlov, and A. N. Semenov. *Journal of Statistical Physics* **38**, 161 (1985).
113. B. Neway, Å . Westberg, A. Mattozzi, M. S. Hedenqvist, M. Giacinti Baschetti, V. B. F. Mathot, and U. W. Gedde. *Polymer* **45**, 3913 (2004).
114. H. R. Brown and Th. P. Russell. *Macromolecules* **29**, 798 (1996).
115. P. G. de Gennes. *European Physical Journal E* **2**, 201 (2000).
116. Sh. Xu, P. A. O'Connell, and G. B. McKenna. *Journal of Chemical Physics* **132**, 184902 (2010).
117. K. Nakashima, K. Tsuboi, H. Matsumoto, R. Ishige, M. Tokita, J. Watanabe, and A. Tanioka. *Macromolecular Rapid Communications* **31**, 1641 (2010).
118. A. Arinstein. *Journal of Polymer Science, Part B: Polymer Physics* **51**, 756 (2013).
119. E. Helfand and G. H. Fredrickson. *Physical Review Letters* **62**, 2468 (1989).

120. Sh. Saito, K. Matsuzaka, and T. Hashimoto. *Macromolecules* **32**, 4879 (1999).
121. I. Morfin, P. Lindner, and F. Boué. *European Physical Journal E* **15**, 41 (2004).
122. C. Teko ĝ lu, L. J. Gibson, T. Pardoen, and P. R. Onck. *Progress in Materials Science* **56**, 109 (2001).
123. Z. Hashin. *Journal of Applied Mechanics* **29**, 143 (1962).
124. R. Hill. *Journal of Mechanics and Physics of Solids* **13**, 213 (1965).
125. A. Arinstein, Y. Liu, M. Rafailovich, and E. Zussman. *Europhysics Letters* **93**, 46001 (2011).
126. M. Alcoutlabi and G. B. McKenna. *Journal of Physics: Condensed Matter* **17**, R461 (2005).
127. A. K. Oultache, X. Kong, Ch. Pellerin, J. Brisson, M. P ė zolet, and R. E. Prud'homme. *Polymer* **42**, 9051 (2001).
128. J. J. Iruin, G. M. Guzman, and J. Ansorena. *Die Makromolekulare Chemie* **182**, 2789 (1981).
129. G. B. McKenna. *European Physical Journal–Special Topics* **141**, 291 (2007).
130. M. T. McDowell, A. M. Leach, and K. Gall. *Modelling and Simulation in Materials Science and Engineering* **16**, 045003 (2008).
131. T. Yoshioka, R. Dersch, M. Tsuji, and A. K. Schaper. *Polymer* **51**, 2383 (2010).
132. M. V. Kakade, S. Givens, K. Gardner, K. H. Lee, D. B. Chase, and J. F. Rabolt. *Journal of the American Chemical Society* **129**, 2777 (2007).
133. L. M. Bellan and H. G. Craighead. *Polymer* **49**, 3125 (2008).
134. M. Richard-Lacroix and C. Pellerin. *Macromolecules* **45**, 1946 (2012).
135. M. Richard-Lacroix and C. Pellerin. *Macromolecules* **46**, 9473 (2013).
136. D. Alhazov, A. Gradys, P. Sajkiewicz, A. Arinstein, and E. Zussman. *European Polymer Journal* **49**, 3851 (2013).
137. D. Alhazov, M. Burman, A. Arinstein, and E. Zussman. *Journal of Polymer Science, Part B: Polymer Physics* **53**, 1254 (2015).
138. S. Deng, A. Arinstein, Y. Cohen, and E. Zussman. *Journal of Polymer Science, Part B: Polymer Physics* **55**, 506 (2017).

Index